NMR 29

Basic Principles and Progress

Editors: P. Diehl E. Fluck H. Günther
R. Kosfeld J. Seelig

Special Applications

With contributions by
A. Bunn, N. J. Clayden, D. Freude, J. Haase

With 55 Figures

Springer-Verlag
Berlin Heidelberg New York
London Paris Tokyo
Hong Kong Barcelona Budapest

ISBN 978-3-642-50048-0 ISBN 978-3-642-50046-6 (eBook)
DOI 10.1007/978-3-642-50046-6

Softcover reprint of the hardcover 1st edition 1973

Typesetting: Thomson Press (India) Ltd, New Delhi

51/3020 - 5 4 3 2 1 0 Printed on acid-free paper

Guest-Editors

Professor Dr. Harry Pfeifer
Fachbereich Physik, Universität Leipzig
Linnéstr. 5, 04103, Leipzig, FRG

Dr. Philip Barker
BHP Research & New Technology, Melbourne Research Laboratories
P.O. Box 264, Clayton, Victoria 3168, Australia 3170

Table of Contents

Quadrupole Effects in Solid-State Nuclear Magnetic Resonance

D. Freude and J. Haase

Fachbereich Physik der Universität Leipzig, O-7010 Leipzig, FRG

Table of Contents

NMR Basic Principles and Progress, Vol. 29

During the last decade there was a growing interest in high-resolution solid state nuclear magnetic resonance (NMR) of quadrupole nuclei, due in part, to the availability of higher magnetic fields and the demand for characterization of inorganic materials. As a consequence, experimental techniques traditionally used in the study of spin-1/2 nuclei have been adapted for use with quadrupole nuclei, e.g., sample spinning, nutation spectroscopy, two-pulse free induction decay and echo methods. This review presents the basic theory and a discussion of experimental techniques for NMR studies of nuclei with half-integer spins in powder samples, including the effect of quadrupole nuclei on the MAS NMR lineshape of spin-1/2 nuclei. A survey of quadrupole parameters determined by means of NMR, published after 1982, is contained in the last section.

Preface

The applicability of nuclear magnetic resonance (NMR) spectroscopy continues to expand in physics, chemistry, material science, geology, biology, and medicine. An increasing range of NMR techniques is employed for either spectroscopic studies or imaging purposes, most becoming possible because of the high magnetic fields of superconducting magnets (almost an order of magnitude higher than ordinary iron magnets), and the availability of fast computers for controlling purposes and final data processing. These improvements together with the realization of fast mechanical sample spinning and new pulse techniques have led to high-resolution solid-state nuclear magnetic resonance. Because the perturbing effect of the electric quadrupole interaction becomes less important at higher magnetic fields, the investigation of powder samples with NMR techniques has become more feasible. As a result, the number of publications concerning NMR studies of quadrupole nuclei in powder samples rapidly increased in the eighties. However, since 1987 this number of publications is rather stable and amounts to about 100 per year.

In this review we briefly describe the theory necessary for the understanding of the most often applied techniques. Basic formulas are taken from well known monographs, e.g. Abragam [1, 2], however, SI-units are used throughout this review. The problem of excitation is discussed in more detail, after presenting the basic theory. More recently developed techniques for the study of quadrupole nuclei with half-integer spins, will then be discussed. For integer spins, especially the solid-state deuterium magnetic resonance, we refer to H. W. Spiess [3]. Studies of single crystals will not be considered, as well. In vol. 3 of this series (NMR-Basic Principles and Progress), published in 1971, O. Kanert and M. Mehring presented a review of "Static quadrupole effects in disordered cubic solids" [4]. Also, we would like to mention the "classic" review of M. H. Cohen and F. Reif of "Quadrupole effects in NMR studies of solids" [5] and the review about "Multipole NMR" of B. C. Sanctuary and T. K. Halstead [6], an approach where the density matrix is represented by its irreducible components.

The state of the art in NQR and zero-field NMR is described by Zax [7] and in the Proceedings of the Eight International Symposium on NQR Spectroscopy [8], the basic principles were explained by T. P. Das and E. L. Hahn [9]. For SQUID (superconducting quantum interference device) detected nuclear resonance we refer to the original papers [10–15]. In this review, pure NQR will not be discussed.

A survey of nuclear quadrupole frequency data published before the end of 1982 is given by H. Chihara and N. Nakamura in Landolt-Börnstein, vol. 20 [16]. Values of the chemical shift of quadrupole nuclei in solids can be found in books such as "Multinuclear NMR" edited by J. Mason [17]. We present in Sect. 9 some electric field gradient and chemical shift data published from 1983 to 1992, in tables for the nuclei ^{27}Al, ^{23}Na, and ^{17}O and a few recent references for other quadrupole nuclei with half-integer spins.

In early investigations, particularly in the case of ionic solids, the attempt was made to interpret the electric field gradient as due to point charges around the ion under study. In this approach it was necessary to take into consideration the Sternheimer factor γ_∞, which accounts for the antishielding of the point charge contributions and of the nuclear quadrupole moment by the core electrons of the ion. The work in this field began with Sternheimer's first paper [18] and was reviewed by himself in 1986 [19]. Sternheimer factors which include self-consistency and solid-state effects were calculated by P. C. Schmidt et al. [20] (see also Sen et al. [21] for rare earth atoms) using the Watson sphere model [22]. Metallic and semi-metallic systems, which will not be considered in this review, require a different theory to interpret the nuclear quadrupole interaction, as described by T. P. Das and P. S. Schmidt [23]. In the 1980s there was a growing interest in first-principles description of solids, which, concerning the electric field gradient, do not rely on any Sternheimer antishielding factor. A new approach by P. Blaha [24] yields satisfactory results for the electric field gradient in high temperature superconductors. For silicates, ab initio calculations of model clusters consisting of two or three silicon atoms can be used in order to obtain ^{17}O NMR parameters [25]. The comparison of experimental results with the calculated electric field gradient may yield insight concerning bonding effects [26] or confirm assumptions made in quantum chemical methods.

Multiple-quantum (MQ) NMR (cf. Ernst et al. [27]) has received attention in studies of quadrupole nuclei for two reasons: first, some multi-quantum transitions are not influenced by the quadrupole interaction, second, the relaxation behavior of the transitions yields additional information about the system under study. The majority of studies are devoted to the spin-1 nuclei ^{2}H and ^{14}N (cf. Ernst et al. [27] and references therein). The first experiments on half-integer quadrupole nuclei were performed on ^{27}Al in single crystals of Al_2O_3 by Hatanaka et al. [28] and on ^{23}Na in single crystals of sodium ammonium tartrate tetrahydrate by S. Vega [29]. Kowalewski et al. [30] observed double and triple-quantum coherences for ^{7}Li in a sample of macroscopically oriented LiDNA using nonselective pulses. Man [31] studied a single crystal of $LiTaO_3$ and showed that certain combinations of spin-lock and rotary echo detect the double-quantum coherence and the magnetization independently. Furó et al. [32, 33] demonstrated the use of multiple-quantum filter and 2D echo techniques to reduce the overlap of signals of quadrupole nuclei in anisotropic systems. Rooney et al. [34] measured double-quantum spectra of ^{23}Na in unoriented lyotropic liquid crystals. The observation of MQ transitions for half-integer quadrupole nuclei in typical powder samples is spars: Van der Maarel [35] performed echo train experiments on ^{23}Na in sodium poly-(methylacrylate) ion-exchange resin and monitored triple-quantum coherences by applying an additional coherence transfer pulse after the pulse train. Nielsen, Bildsoe and Jakobsen [36, 37] introduced a MQ MAS version of the 2D nutation experiment and demonstrated it with ^{27}Al and ^{23}Na spectra of $Ca_3Al_2O_6 \cdot 6H_2O$ and $NaNO_3$ powder, respectively, cf. Sect. 4.

Magnetic resonance imaging (MRI) techniques are now being used to investigate solids. The most recent advances are described by P. Jezzard et al. [38]. Only two of the original publications were concerned with quadrupole nuclei: Suids et al. used convolution back projection to obtain 2D ^{23}Na images of ion motion in a solid conductor (Na_{1-x}, K_x) β-alumina single crystal [39] and demonstrated the spatial variation of ^{23}Na signal intensity in an NaCl crystal which results from an impact-induced inhomogeneous defect distribution [40].

1 Basic Theory

1.1 Interaction of Nuclei with External Magnetic Fields

The Hamiltonian for a nuclear spin $\boldsymbol{I}$, interacting with an external magnetic field $\boldsymbol{B}$ is

$$\mathscr{H} = -\gamma\hbar\boldsymbol{B}\cdot\boldsymbol{I}, \tag{1.01}$$

where $h = 2\pi\hbar$ denotes Planck's constant and γ the gyromagnetic ratio, cf. Abragam [1]. For the case of a static external magnetic field B_0 pointing in z-direction of the laboratory frame we have

$$\mathscr{H}_L = -\gamma\hbar I_z B_0, \tag{1.02}$$

and, with the definition of the Larmor frequency ($\nu_L = \omega_L/2\pi$),

$$\boldsymbol{\omega}_L = -\gamma\boldsymbol{B}_0, \tag{1.03}$$

Eq. (1.02) can be rewritten as

$$\mathscr{H}_L = \hbar\omega_L I_z. \tag{1.04}$$

Application of a radio frequency field in y-direction,

$$B_y(t) = 2\cdot B_{rf}\cos(\omega t), \tag{1.05}$$

enlarges the Hamiltonian

$$\mathscr{H} = \hbar\omega_L I_z + 2\hbar\omega_{rf}\cos(\omega t) I_y, \tag{1.06}$$

where ω_{rf} is defined as

$$\omega_{rf} = \gamma B_{rf}. \tag{1.07}$$

1.2 Irreducible Tensor Operator Calculus

It is useful to express the internal interactions of a nuclear spin in the notation of irreducible tensor operators, cf. Weissbluth [42]. If $T_q^{(k)}$ denotes one of the

$(2k+1)$ components of an irreducible tensor operator of rank k, then, it must under a coordinate rotation, $\boldsymbol{r}' = R\boldsymbol{r}$, transform as

$$P_R T_q^{(k)} P_R^{-1} = \sum_{q'} T_{q'}^{(k)} D_{q'q}^{(k)}(R), \qquad q, q' = k, k-1, \ldots, -k. \tag{1.08}$$

where $D_{q'q}^{(k)}(R)$ denotes the matrix elements of the irreducible representation $D^{(k)}$ of the group of the ordinary three-dimensional rotations. The transformation operator P_R is given by

$$P_R = \exp\{i\Omega \boldsymbol{n}\cdot\boldsymbol{J}\}, \tag{1.09}$$

where $\boldsymbol{J}$ is the total angular momentum operator, $\boldsymbol{n}$ is the unit vector pointing in the direction of rotation and Ω represents the angle of rotation. Such a rotation transforms the eigenfunctions, $|\mathrm{Im}\rangle$, of the angular momentum operator I_z of a nuclear spin $\boldsymbol{I}$, as

$$P_R|\mathrm{Im}\rangle = \sum_{m'} |\mathrm{Im}'\rangle D_{m'm}^{(I)}(R), \tag{1.10}$$

where

$$D_{m'm}^{(k)}(\alpha, \beta, \gamma) = \exp\{i\alpha m'\}\, d_{m'm}^{(k)}(\beta) \exp\{i\gamma m\},$$

with values of $d_{m'm}^{(k)}(\beta)$ given in Table 1.1 for $k = 1/2$, 1, 3/2, 2, and 5/2 [41]. Values for $k = 4$ can be found in Ref. [43]. As presented in Fig. 1.1, a positive rotation to a frame (x, y, z) about the Euler angles includes the rotation α about the original z axis, the rotation β about the obtained y′ axis, and the rotation γ about the final z″ axis (cf. Rose [44]). An equivalent definition of Eq. (1.08) of an irreducible tensor operator is given by the commutators [42]

$$[J_0, T_q^{(k)}] = qT_q^{(k)}, \qquad [J_{\pm 1}, T_q^{(k)}] = \mp\sqrt{\{k(k+1) - q(q \pm 1)\}/2}\, T_{q\pm 1}^{(k)}, \tag{1.11}$$

where J denotes the total angular momentum operator and $J_0 = J_z$, $J_{\pm 1} = \mp \dfrac{J_\pm}{\sqrt{2}}$.

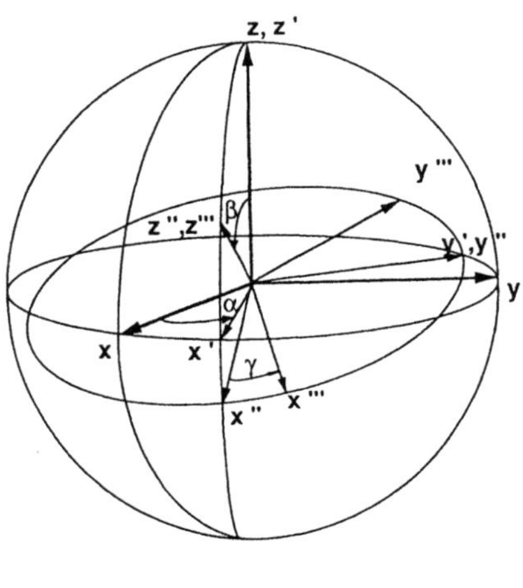

Fig. 1.1. Definition of the Euler angles: A positive rotation to a frame (x, y, z) about the Euler angles includes the rotation α about the original z axis, the rotation β about the obtained y′ axis, and the rotation γ about the final z″ axis

Table 1.1. The reduced real rotation matrices $d^{(k)}_{q'q}(\beta)$ as compiled by I. Wolf [41] for rank $k = 1/2$, 1, 3/2, 2, and 5/2. The operators $D^{(k)}_{q'q}(\alpha, \beta, \gamma) = \exp\{i\alpha q'\} d^{(k)}_{q'q}(\beta) \exp\{i\gamma q\}$, cf. Eq. (1.08), are matrix elements of the irreducible representations of three-dimensional rotation group $O^+(3)$

$k=1/2$	$q'\backslash q$	$-1/2$	$1/2$					
	$-1/2$	A_+	$-A_-$					$A_\pm = \sqrt{\frac{1\pm\cos\beta}{2}}$
	$1/2$	A_-	A_+					
$k=1$	$q'\backslash q$	-1	0	1				$B_\pm = A_\pm^2$
	-1	B_+	$-C$	B_-				
	0	C	D	$-C$				$C = \sqrt{1/2}\sin\beta$
	1	B_-	C	B_+				$D = \cos\beta$
$k=3/2$	$q'\backslash q$	$-3/2$	$-1/2$	$1/2$	$3/2$			
	$-3/2$	E_+	$-F_+$	F_-	$-E_-$			$E_\pm = A_\pm^3$
	$-1/2$	F_+	G_+	$-G_-$	F_-			$F_\pm = \sqrt{3/2}CA_\pm$
	$1/2$	F_-	G_-	G_+	$-F_+$			$G_\pm = (1/2)(3\cos\beta \mp 1)A_\pm$
	$3/2$	E_-	F_-	F_+	E_+			
$k=2$	$q'\backslash q$	-2	-1	0	1	2		$H_\pm = A_\pm^4$
	-2	H_+	$-I_+$	K	$-I_-$	H_+		$I_\pm = \sqrt{2}CA_\pm^2$
	-1	I_+	L_+	$-M$	L_-	$-I_-$		$K = \sqrt{3/2}C^2$
	0	K	M	N	$-M$	K		$L_\pm = (2\cos\beta \mp 1)A_\pm^2$
	1	I_-	L_-	M	L_+	$-I_+$		$M = \sqrt{3}CD$
	2	H_-	I_-	K	I_+	H_+		$N = (1/2)(3\cos^2\beta - 1)$
$k=5/2$	$q'\backslash q$	$-5/2$	$-3/2$	$-1/2$	$1/2$	$3/2$	$5/2$	
	$-5/2$	O_+	$-P_+$	Q_+	$-Q_-$	P_-	$-O_-$	$O_\pm = A_\pm^5$
	$-3/2$	P_+	R_+	$-S_+$	S_-	R_-	P_-	$P_\pm = \sqrt{5/2}CA_\pm^3$
	$-1/2$	Q_+	S_+	T_+	$-T_-$	S_-	$-Q_-$	$Q_\pm = \sqrt{5/2}C^2A_\pm$
	$1/2$	Q_-	$-S_-$	T_-	T_+	$-S_+$	Q_+	$R_\pm = A_\pm^5 - 2C^2A_\pm$
	$3/2$	P_-	$-R_-$	$-S_-$	S_+	R_+	$-P_+$	$S_\pm = 2CA_\pm^3 - \sqrt{9/2}C^2A_\mp$
	$5/2$	O_-	P_-	Q_-	Q_+	P_+	O_+	$T_\pm = A_\pm^5 - 3C^2A_\pm - \sqrt{9/2}CA_\mp^3$

If only the nuclear spin $\boldsymbol{I}$ is concerned, we have

$$[I_0, T^{(k)}_q] = qT^{(k)}_q, \quad [I_{\pm 1}, T^{(k)}_q] = \mp\sqrt{\{k(k-1) - q(q\pm 1)\}/2}\,T^{(k)}_{q\pm 1},$$

with

$$I_0 = I_z, \quad I_{\pm 1} = \mp\frac{I_\pm}{\sqrt{2}}, \quad \text{and} \quad I_\pm = I_x \pm iI_y. \tag{1.12}$$

The tensor product, $U^{(K)}_Q$, of two operators with the components $T^{(K)}_q$ and $V^{(k')}_{q'}$ is defined as [42]

$$U^{(K)}_Q = \sum_{q,q'} T^{(k')}_q V^{(k')}_{q'} \langle kk'qq' | kk'KQ\rangle. \tag{1.13}$$

$\langle kk'qq'|kk'KQ\rangle$ are Clebsch–Gordan coefficients. $U_Q^{(K)}$ is an irreducible tensor operator. $U_0^{(0)}$ is the scalar component of that tensor operator and, with the properties of the Clebsch–Gordan coefficients, it is found that

$$U_0^{(0)} = \frac{(-1)^k}{\sqrt{2k+1}} \sum_q (-1)^q T_q^{(k)} V_{-q}^{(k)}, \quad q = -k, -k+1, \ldots, k. \tag{1.14}$$

Most interactions in the NMR can be written in this form [45]. It is important that the operators $T_q^{(k)}$ and $V_q^{(k)}$ often act on different non-interacting systems, the nuclear spin coordinates and spatial coordinates (lattice parameters), respectively. Irreducible tensor calculus is most useful when coordinate transformations have to be considered as in the case of the transformation to the principle axis system (PAS) of a tensor (e.g., for the description of sample spinning).

1.3 Electric Quadrupole Interaction

In the principle axis system (X, Y, Z) of the electric field gradient tensor the Hamiltonian of the electric quadrupole interaction can be written as, cf. Abragam [1] p. 166,

$$\mathscr{H}_Q = \frac{e^2qQ}{4I(2I-1)}\{3I_Z^2 - I(I+1) + \tfrac{1}{2}\eta(I_+^2 + I_-^2)\}, \tag{1.15}$$

with the definitions, $eq = V_{ZZ}$, and $\eta = \dfrac{V_{XX} - V_{YY}}{V_{ZZ}}$. It is the convention to define the components of the traceless tensor $|V_{ZZ}| \geqq |V_{YY}| \geqq |V_{XX}|$. The value q alone has no physical meaning in SI units. Q is the quadrupole moment, when multiplied by the elementary charge e, eQ is called the electric quadrupole moment. Here, $I_\pm = I_X \pm iI_Y$ refers to the principle axis system, as well. The quadrupole coupling constant, C_{qcc}, is commonly defined as

$$C_{qcc} = \frac{e^2qQ}{h},$$

whereas, for the quadrupole frequency, ω_Q or ν_Q, different definitions exist in the literature. We use

$$\omega_Q = \frac{3e^2qQ}{2I(2I-1)\hbar}. \tag{1.16}$$

in agreement with Abragam [1] and the well-established definition for half-integer spin nuclei in the field of NQR. For different definitions cf. Sect. 4.

The quadrupole interaction can be regarded as a scalar product of two irreducible tensor operators of rank two [45], the tensor of the electric field gradient and the quadrupole tensor, cf. Eq. (1.14). Using this notation the quadrupole interaction is described by

$$\mathscr{H}_Q = C_Q \sum_{q=-2}^{+2} (-1)^q T_q^{(2)} V_{-q}^{(2)}, \tag{1.17}$$

where

$$T_0^{(2)} = \frac{1}{\sqrt{6}}(3I_z^2 - I(I+1)), \quad T_{\pm 1}^{(2)} = \frac{1}{\sqrt{2}}(I_{\pm 1} I_z + I_z I_{\pm 1}), \quad T_{\pm 2}^{(2)} = (I_{\pm 1})^2$$

$$C_Q = \frac{eQ}{2I(2I-1)}, \quad I_{\pm 1} = \mp \frac{I_\pm}{\sqrt{2}}.$$

In the principle axis system (PAS) of the tensor of the electric field gradient we have for $V_q^{(2)}$

$$V_{\mathrm{PAS}\,0}^{(2)} = \sqrt{\frac{3}{2}} V_{ZZ}, \quad V_{\mathrm{PAS}\,\pm 1}^{(2)} = 0, \quad V_{\mathrm{PAS}\,\pm 2}^{(2)} = \frac{1}{2} V_{ZZ} \eta. \tag{1.18}$$

In an arbitrary frame (x, y, z) which can be brought into coincidence with the principle axis system by the three successive rotations: γ about the z-axis, β about the obtained y′-axis, α about the Z-axis, we obtain from Eq. (1.08) with the help of Table 1.1 for the new $V_0^{(2)}$ component

$$V_0^{(2)} = \sqrt{\frac{3}{2}} V_{ZZ} \left\{ \frac{3\cos^2\beta - 1}{2} + \frac{\eta}{2} \sin^2\beta \cos 2\alpha \right\}, \tag{1.19}$$

which is needed for the part of $\mathscr{H}_Q$ which commutes with I_z (the Zeeman Hamiltonian). This part is denoted by

$$\begin{aligned} \mathscr{H}_Q^{(0)} &= C_Q T_0^{(2)} V_0^{(2)} \\ &= \frac{eQV_{ZZ}}{4I(2I-1)} (3I_z^2 - I(I+1)) \left\{ \frac{3\cos^2\beta - 1}{2} + \frac{\eta}{2} \sin^2\beta \cos 2\alpha \right\}. \end{aligned} \tag{1.20}$$

By substituting $V_{ZZ} = eq$ and using Eq. (1.16) we find

$$\mathscr{H}_Q^{(0)} = \frac{\hbar\omega_Q}{6} \left\{ \frac{3\cos^2\beta - 1}{2} + \frac{\eta}{2} \sin^2\beta \cos 2\alpha \right\} (3I_z^2 - I(I+1)). \tag{1.21}$$

With the angular dependent quadrupole frequency

$$\omega_Q' = \omega_Q \left\{ \frac{3\cos^2\beta - 1}{2} + \frac{\eta}{2} \sin^2\beta \cos 2\alpha \right\} \tag{1.22}$$

one can write

$$\mathscr{H}_Q^{(0)} = \frac{\hbar\omega_Q'}{6} (3I_z^2 - I(I+1)). \tag{1.23}$$

Using the relations between Cartesian tensors and irreducible tensors, it is possible to express the spherical components in an arbitrary frame as [42]

$$V_0^{(2)} = \sqrt{\frac{3}{2}} V_{zz}, \quad V_{\pm 1}^{(2)} = \mp (V_{zx} \pm \mathrm{i} V_{zy}), \quad V_{\pm 2}^{(2)} = \frac{1}{2}(V_{xx} - V_{yy}) \pm \mathrm{i} V_{xy}. \tag{1.24}$$

1.4 Magnetic Dipole Interaction Between Nuclei

The Hamiltonian, describing the interaction between two magnetic moments with the distance $\boldsymbol{r}_{ik}$ is [1]

$$\mathscr{H}_D = -\frac{\mu_0}{4\pi} \frac{\gamma_i \gamma_k \hbar^2}{r_{ik}^3} \left\{ \frac{3(\boldsymbol{I}_i \cdot \boldsymbol{r}_{ik}) \cdot (\boldsymbol{I}_k \cdot \boldsymbol{r}_{ik})}{r_{ik}^2} - \boldsymbol{I}_i \cdot \boldsymbol{I}_k \right\}. \tag{1.25}$$

Using the irreducible tensor notation one may write [45, 46]

$$\mathscr{H}_D = C_D^{ik} \sum_{q=-2}^{+2} (-1)^q T_q^{(2)} V_{-q}^{(2)}, \tag{1.26}$$

where

$$T_0^{(2)} = \frac{1}{\sqrt{6}}(3 I_{z,i} I_{z,k} - \boldsymbol{I}_i \cdot \boldsymbol{I}_k), \quad T_{\pm 1}^{(2)} = \frac{1}{\sqrt{2}}(I_{\pm 1,i} I_{z,k} + I_{z,i} I_{\pm 1,k}),$$

$$T_{\pm 2}^{(2)} = I_{\pm 1,i} I_{\pm 1,k},$$

$$I_{\pm 1} = \mp \frac{I_\pm}{\sqrt{2}}, \quad C_D^{ik} = -\frac{\mu_0}{2\pi} \gamma_i \gamma_k \hbar^2, \quad V_0^{(2)} = \sqrt{\frac{3}{2}} r_{ik}^{-3},$$

$$V_{\pm 1}^{(2)} = 0, \quad V_{\pm 2}^{(2)} = 0.$$

For **homonuclear dipolar interaction**, $\gamma_i = \gamma_k$, the part of $\mathscr{H}_D$, commuting with $I_{z,i} + I_{z,k}$, in analogy to Eq. (1.19) and with Eq. (1.26) for a pair (i, k) of spins with the distance $\boldsymbol{r}_{ik}$, is [1]

$$\mathscr{H}_D^{(0)} = -\frac{\mu_0}{4\pi} \gamma^2 \hbar^2 \left\{ \frac{3 \cos^2 \beta_{ik} - 1}{2 r_{ik}^3} \right\} (3 I_{z,i} I_{z,k} - \boldsymbol{I}_i \cdot \boldsymbol{I}_k). \tag{1.27}$$

For **heteronuclear dipolar interaction**, $\gamma_i \neq \gamma_k$, the part of $\mathscr{H}_D$, commuting with $\gamma_i I_{z,i} + \gamma_k I_{z,k}$, is [1]

$$\mathscr{H}_D^{(0)} = -\frac{\mu_0}{2\pi} \gamma_i \gamma_k \hbar^2 \left\{ \frac{3 \cos^2 \beta_{ik} - 1}{2 r_{ik}^3} \right\} I_{z,i} I_{z,k}. \tag{1.28}$$

1.5 Anisotropy of the Chemical Shift

The Hamiltonian describing the anisotropy of the shielding is, cf. Mehring [46],

$$\mathscr{H}_{CSA} = C_{CSA} \left\{ T_0^{(0)} V_0^{(0)} + \sum_{q=-2}^{+2} (-1)^q T_q^{(2)} V_{-q}^{(2)} \right\} \tag{1.29}$$

with

$$T_0^{(0)} = -\frac{1}{\sqrt{3}} I_0 B_0, \quad T_0^{(2)} = \sqrt{\frac{2}{3}} I_0 B_0, \quad T_{\pm 1}^{(2)} = \frac{1}{\sqrt{2}} I_{\pm 1} B_0, \quad T_{\pm 2}^{(2)} = 0,$$

$$I_{\pm 1} = \mp \frac{I_\pm}{\sqrt{2}}, \qquad C_{\mathrm{CSA}} = \gamma\hbar,$$

$$V_0^{(0)} = -\frac{1}{\sqrt{3}}(\sigma_{XX} + \sigma_{YY} + \sigma_{ZZ}) \equiv -\sqrt{3}\sigma_{\mathrm{iso}},$$

$$V_0^{(2)} = \sqrt{\frac{3}{2}}(\sigma_{ZZ} - \sigma_{\mathrm{iso}}) \equiv \sqrt{\frac{3}{2}}\delta,$$

$$V_{\pm 2}^{(2)} = \frac{1}{2}\eta\delta \equiv \frac{1}{2}(\sigma_{XX} - \sigma_{YY}), \quad V_{\pm 1}^{(2)} = 0.$$

$\sigma_{XX}, \sigma_{YY}, \sigma_{ZZ}$ denote the components of the shielding tensor in the principle axis system.

1.6 Density Operator Formalism—Interaction Representation

For the calculation of the free induction decay it is most useful to transform the equations of motion of the spin system into the so-called *interaction representation*, by separating the Hamiltonian $\mathscr{H}_0 = \hbar\omega I_z$, where ω is the carrier frequency of the irradiated pulse, since it is equivalent to the calculation of the demodulated free induction decay.

The total Hamiltonian of the system, in the Schrödinger representation, can be written as $\mathscr{H}(t) = \mathscr{H}_0 + \mathscr{H}_1(t)$. By going from the Schrödinger representation (*s*) to the interaction representation (*i*) with respect to $\mathscr{H}_0$, the wave function $|\Phi_s\rangle$ and any operator A_s have to be transformed by the unitary operator

$$P_i = \exp\left\{\frac{i}{\hbar}\mathscr{H}_0(t - t_0)\right\},$$

so that [47]

$$|\Phi_i\rangle = P_i|\Phi_s\rangle, \quad A_i(t) = P_i A_s P_i^{-1}. \tag{1.30}$$

The equations of motion for the wave function and the density operator ρ_i in the interaction representation are [47]

$$\frac{\partial}{\partial t}|\Phi_i\rangle = -\frac{i}{\hbar}\mathscr{H}_{1,i}|\Phi_i\rangle, \quad \frac{\partial}{\partial t}\rho_i = -\frac{i}{\hbar}[\mathscr{H}_{1,i}, \rho_i]. \tag{1.31}$$

Using $\mathscr{H}_0 = \hbar\omega I_z$, and the resonance offset $\Delta\omega = \omega_L - \omega$, we have for the Hamiltonian in the interaction representation, according to Eq. (1.30),

$$\mathscr{H}_{1,i} = \hbar\Delta\omega I_z + \exp\{i\omega I_z t\}\mathscr{H}_1 \exp\{-i\omega I_z t\} \tag{1.32}$$

A formal solution to the equation of motion of the density operator, Eq. (1.31), is obtained by the unitary evolution operator $U_i(t, t_0)$

$$\rho_i(t) = U_i(t, t_0)\rho_i(t_0)U_i(t, t_0)^{-1}. \tag{1.33}$$

The evolution operator $U_i(t, t_0)$ obeys the equation

$$\frac{\partial}{\partial t}U_i(t, t_0) = -\frac{i}{\hbar}\mathcal{H}_{1,i}U_i(t, t_0). \tag{1.34}$$

If the operator $\mathcal{H}_{1,i}$ is time-independent the solution is

$$U_i(t, t_0) = \exp\left\{-\frac{i}{\hbar}\mathcal{H}_{1,i}(t - t_0)\right\}. \tag{1.35}$$

The general solution to Eq. (1.34) is

$$U_i(t, t_0) = 1 - \frac{i}{\hbar}\int_{t_0}^{t}\mathcal{H}_{1,i}(t_1)U_i(t_1, t_0)\,dt_1, \tag{1.36}$$

$$U_i(t, t_0) = 1 - \frac{i}{\hbar}\int_{t_0}^{t}\mathcal{H}_{1,i}(t_1)\,dt_1 + (-i/\hbar)^2\int_{t_0}^{t}\mathcal{H}_{1,i}(t_1)\,dt_1\int_{t_0}^{t_1}\mathcal{H}_{1,i}(t_2)\,dt_2 + \cdots \tag{1.37}$$

or, with the help of the time ordering operator T [46]

$$U_i(t, t_0) = T\exp\left\{-\frac{i}{\hbar}\int_{t_0}^{t}\mathcal{H}_{1,i}(t_1)\,dt_1\right\}. \tag{1.38}$$

The Magnus expansion, cf. Haeberlen [45], can be used in order to describe the evolution in Eq. (1.38) for a time-dependent, but periodic Hamiltonian, $\mathcal{H}_{1,i}(t + Nt_c) = \mathcal{H}_{1,i}(t)$ with integer N

$$U_i(t_c, t_0 = 0) = \exp\left\{-\frac{i}{\hbar}t_c(\overline{\mathcal{H}^{(0)}} + \overline{\mathcal{H}^{(1)}} + \overline{\mathcal{H}^{(2)}} + \cdots)\right\}, \tag{1.39}$$

where

$$\overline{\mathcal{H}^{(0)}} = \frac{1}{t_c}\int_0^{t_c}\mathcal{H}_{1,i}(t)\,dt, \quad \overline{\mathcal{H}^{(1)}} = \frac{-i}{2t_c}\int_0^{t_c}dt_2\int_0^{t_2}dt_1[\mathcal{H}_{1,i}(t_2), \mathcal{H}_{1,i}(t_1)],$$

and the odd terms disappear for $\mathcal{H}_{1,i}(t_c - t) = \mathcal{H}_{1,i}(t)$. For higher order terms cf. Ref. [45].

The superoperator notation, cf. Jeener [48], which considers the evolution of an operator as a mapping in a new space, the Liouville space (with operators as basis vectors and new superoperators as operators, and an appropriately defined scalar product). In such a notation, the mapping of the density operator, $\rho(t_0) \rightarrow \rho(t)$, is described by the action of a superoperator $\hat{U}(t, t_0)$,

$$\hat{U}(t, t_0)\rho(t_0) \equiv U(t, t_0)\rho(t_0)U(t, t_0)^{-1}, \tag{1.40}$$

and, the Hamiltonian by a Liouville operator, $\hat{H}$,

$$\frac{\partial}{\partial t}\rho_i = -\frac{i}{\hbar}\hat{H}_{1,i}\rho_i \equiv -\frac{i}{\hbar}[\mathscr{H}_{1,i}, \rho_i]. \tag{1.41}$$

Ordinary operators are regarded as vectors in the Liouville space

$$\hat{U}(t, t_0)|\rho(t_0)\rangle = |\rho(t)\rangle, \quad \frac{\partial}{\partial t}|\rho_i\rangle = -\frac{i}{\hbar}\hat{H}_{1,i}|\rho_i\rangle. \tag{1.42}$$

The scalar product of two vectors in the Liouville space is defined by

$$\langle A|B\rangle \equiv \mathrm{tr}\,\{A^\dagger B\}, \tag{1.43}$$

where the cross denotes the adjoint operator.

1.7 Calculation of the Free Induction Decay (FID)

The expectation value of any self-adjoint operator, $O = O^\dagger$, is

$$\langle O\rangle = \frac{\mathrm{tr}\,\{\rho O\}}{\mathrm{tr}\,\{\rho\}} \equiv \frac{\langle O|\rho\rangle}{\langle \mathbb{1}|\rho\rangle} \tag{1.44}$$

with $\mathbb{1}$ being the unity operator. In a frame rotating at ω about the z-axis of the laboratory system the observable O transforms according to Eqs. (1.08) and (1.09)

$$O_r = \exp\{-i\omega I_z t\} O \exp\{i\omega I_z t\}. \tag{1.45}$$

For the expectation value it follows from Eq. (1.44)

$$\langle O_r\rangle = \frac{\mathrm{tr}\,\{\rho O_r\}}{\mathrm{tr}\,\{\rho\}} = \frac{\mathrm{tr}\,\{\rho_i O\}}{\mathrm{tr}\,\{\rho\}}, \tag{1.46}$$

where ρ_i is the density operator in the interaction representation.

By means of the phase sensitive quadrature detection, the demodulated signal (FID) is digitized into two sets of data where a unique time difference between the two sets corresponds to a 90° phase shift of the carrier frequency of the pulse. Thus, the FID is a complex voltage and can be calculated by [1, 46]

$$G(t) = C\frac{\mathrm{tr}\,\{\rho_i(t) I_+\}}{\mathrm{tr}\,\{\rho\}} \equiv C\frac{\langle I_-|\rho_i(t)\rangle}{\langle \mathbb{1}|\rho\rangle}, \tag{1.47}$$

where C is a normalization constant defined below in Eq. (1.49).

In the high-temperature approximation, $\|\mathscr{H}_L\| \ll kT$, for a single spin I the density operator is $\rho_0 = (\mathbb{1} - \beta\hbar\omega_L I_z)/\mathrm{tr}\,\{\mathbb{1}\}$. From Eq. (1.39), the time evolution $|\rho_i(t)\rangle = |\mathbb{1}\rangle + a\hat{U}(t, t_0)|I_z\rangle$ follows, where $a = -\beta\hbar\omega_L/\mathrm{tr}\,\{\mathbb{1}\}$. Since all angular momentum operators are traceless we have

$$G(t) = Ca\frac{\langle I_-|\hat{U}(t, t_0)|I_z\rangle}{\langle \mathbb{1}|\rho\rangle}. \tag{1.48}$$

If $G(t)$ is normalized, requiring its maximum magnitude to be unity,

$$C = \frac{\langle \mathbb{1} | \rho \rangle}{a \langle I_x | I_x \rangle}. \tag{1.49}$$

Therefore, it is convenient for the calculation of the FID to define the density operator in the high-temperature approximation as being proportional to I_z,

$$|\rho_0\rangle = |I_z\rangle, \tag{1.50}$$

and, consider the FID to be given by

$$G(t) = \frac{\mathrm{tr}\,\{\rho_i(t) I_+\}}{\mathrm{tr}\,\{I_x^2\}} \equiv \frac{\langle I_- | \rho_i(t) \rangle}{\langle I_x | I_x \rangle}. \tag{1.51}$$

1.8 Examples of Hamiltonians in the Interaction Representation

In the interaction representation the above defined Hamiltonians have the following form, when transformed by $\mathscr{H}_0 = \hbar\omega I_z$, cf. Eq. (1.30),

Static field in z-direction

$$\mathscr{H}_{L,i} = \hbar \Delta\omega I_z, \quad \text{where} \quad \Delta\omega = \omega_L - \omega. \tag{1.52}$$

Magnetic rf field (y-pulse)

$$\mathscr{H}_{rf,i} = \hbar\omega_{rf}\{I_y(1 + \cos(2\omega t)) + I_x \sin(2\omega t)\}, \tag{1.53}$$

by neglecting the time-dependent part

$$\mathscr{H}_{rf,i} = \hbar\omega_{rf} I_y. \tag{1.54}$$

If ϕ is the phase difference of the pulse with respect to the y-direction, positive for a right-handed screw in the positive direction along the axis of rotation, one has from Eq. (1.09)

$$\mathscr{H}_{rf,i} = \hbar\omega_{rf} \exp\{-i\phi I_z\} I_y \exp\{+i\phi I_z\}. \tag{1.55}$$

Quadrupole interaction

$$\mathscr{H}_{Q,i} = C_Q \sum_{q=-2}^{+2} (-1)^q T_q^{(2)} V_{-q}^{(2)} \exp\{iq\omega t\}. \tag{1.56}$$

Using the Magnus expansion, Eq. (1.39), we obtain the first-order contribution

$$\overline{\mathscr{H}_Q^{(0)}} = \frac{\hbar\omega_Q'}{6}(3I_z^2 - I(I+1)), \tag{1.57}$$

for the definition of ω_Q' cf. Eq. (1.22). From the second-order contribution, $\overline{\mathscr{H}_Q^{(1)}}$, the secular part with respect to I_z is

$$\overline{\mathscr{H}}^{(1)}_{\text{Qsec}} = -\frac{\hbar\omega_Q^2}{9\omega_L}\{2I_z[2I_z^2 - I(I+1) + \tfrac{1}{4}]V_{-1}V_1 + I_z[I_z^2 - I(I+1) + \tfrac{1}{2}]V_{-2}V_2\}, \tag{1.58}$$

and the non-secular term

$$\overline{\mathscr{H}}^{(1)}_{\text{Qn-s}} = -\frac{\hbar\omega_Q\omega'_Q}{12\omega_L}\{(I_-I_z^2 + 2I_zI_-I_z + I_z^2I_-)V_{+1} - (I_+I_z^2 + 2I_zI_+I_z + I_z^2I_+)V_{-1} - (I_-^2I_z + I_zI_-^2)V_{+2} + (I_z^2 + I_zI_+^2)V_{-2}\}, \tag{1.59}$$

where, in order to define ω_Q, the largest principle axis component of the electric field gradient tensor, V_{ZZ}, has been extracted from $V_q^{(2)}$, i.e. $V_q = V_q^{(2)}/V_{ZZ}$. The irreducible tensor components, $V_q^{(2)}$ can be expressed by the principle axis components using Eqs. (1.08), (1.18) and Table 1.1, as

$$V_q^{(2)} = \sum_{q'} D^{(2)}_{q'q}(R)V^{(2)}_{\text{PAS}q'}. \tag{1.60}$$

Chemical Shift Anisotropy

After transformation to the laboratory frame the secular part of the anisotropy of the chemical shift is

$$\mathscr{H}^{(0)}_{\text{CSA}} = \gamma\hbar B_0 I_z\left\{\sigma + \delta\left[\frac{3\cos 2\beta - 1}{2} + \frac{\eta}{2}\sin^2\beta\cos 2\alpha\right]\right\}. \tag{1.61}$$

2 Excitation of Quadrupole Nuclei

The quadrupole interaction causes a first-order frequency shift for the satellite transitions, and often, only a fraction of the satellite transitions can therefore be excited by an rf pulse. A strong quadrupole interaction complicates the NMR experiment for two reasons: first, the quadrupole interaction has to be considered for the Hamiltonian describing the interaction during the irradiation of the pulses. Thus, theoretical expressions become more complicated, and, the visualization of the effect of the pulse as a flipping of the magnetization vector no longer holds. Second, short pulses are necessary in order to excite the satellite transitions for powders and the short rf pulse is finally shaped by the probe circuit, because the quality factor of the probe, $Q = f/\Delta f$, limits the bandwidth.

We assume the following relation between the Hamiltonians describing the several interactions

$$\|\mathscr{H}_L\| \gg \|\mathscr{H}_Q\| \gg \|\mathscr{H}_D\|, \|\mathscr{H}_{\text{CSA}}\|. \tag{2.01}$$

The strength of the interaction of the spins with the irradiated rf field, $\mathscr{H}_{rf}$, is variable. Experimentally, $\mathscr{H}_{rf}$ can be made to be large compared to $\mathscr{H}_D, \mathscr{H}_{CSA}$. But, the quadrupole interaction may easily exceed $\mathscr{H}_{rf}$. Thus, the strength of a quadrupole coupling compared with $\mathscr{H}_{rf}$ has to be considered.

The excitation for the traditional continuous wave, stationary method is more simple than for pulse experiments. Although one irradiates an extremely monochromatic wave in a stationary experiment, as long as the system remains in thermal equilibrium (sufficiently weak amplitude of irradiation so that the spin populations do not change) the excitation can be considered as nonselective. When using pulse methods, the time the system needs for regaining the thermal equilibrium is long compared with the duration of pulses. The amplitude of the rf pulses ensures that the population numbers are changed considerably and one faces a non-equilibrium situation, cf. Abragam [1] p. 40. Also, for larger rf amplitudes in stationary experiments one observes saturation effects which may change the line shape of the NMR signal. For quadrupole nuclei this may affect spin–lattice relaxation measurements as one finds discussed in an early paper by Andrew and Tunstall [49].

Since stationary methods are less important today we discuss pulse methods only, and split the discussion into the parts: spectral density of pulses, resonance offset, nonselective excitation, selective excitation of the central transition only and, partly selective excitation.

2.1 Spectral Density of Pulses

A rectangular pulse with the duration τ and the carrier frequency ω_0 produces a frequency spectrum around ω_0, which can be described with $\Omega = \omega_0 - \omega$ by the Fourier transform

$$f(\Omega) = \int_{-\tau/2}^{+\tau/2} \cos(\Omega t)\, dt = \tau \frac{\sin(\Omega\tau/2)}{\Omega\tau/2}. \tag{2.02}$$

The function $f(\Omega)$ equals τ for $\Omega = 0$ and is zero for integer values of the argument $\Omega\tau/2\pi$. That means that for a rectangular pulse of the duration τ the first node in the frequency spectrum occurs at $\Omega = 2\pi/\tau$, in frequency units at $\nu_{offset} = 1/\tau$. The spectral energy density, $E(\Omega) = f^2(\Omega)$, is proportional to the square of the rf field strength and can be normalized by

$$\int_{-\infty}^{\infty} E(\Omega)\, d\Omega = \frac{1}{\mu_0} B_{rf}^2 = \frac{1}{\mu_0 \gamma^2} \omega_{rf}^2. \tag{2.03}$$

This gives

$$E(\Omega) = \frac{\omega_{rf}^2 \tau}{2\pi\gamma^2\mu_0} \left[\frac{\sin(\Omega\tau/2)}{\Omega\tau/2} \right]^2. \tag{2.04}$$

Since the function in the squared bracket equals 1 for $\Omega \approx 0$, the spectral component of the energy density close to the carrier frequency of the pulse

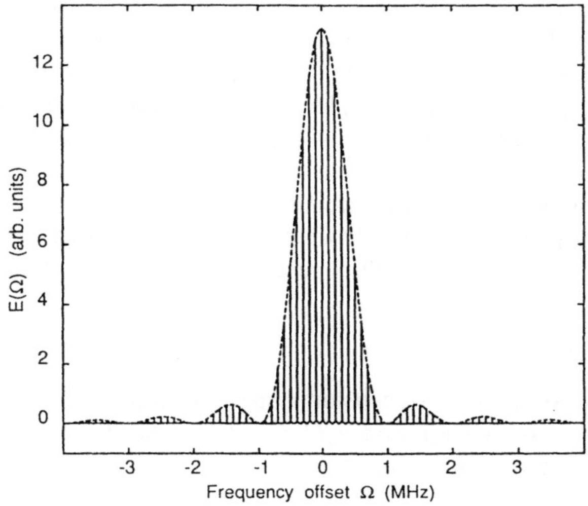

Fig. 2.1. Spectral density of a pulse train consisting of 23 pulses with 1 μs pulse duration and 10 μs pulse distance. The envelope describe a single pulse excitation with one 1 μs-pulse

increases linearly with the pulse duration. Based on Eq. (2.04), we define the usable bandwidth of excitation $\delta\omega_{bw}$ (in analogy to electronics) for a rectangular pulse by

$$E\left(\Omega = \frac{1}{2}\delta\omega_{bw}\right) = \frac{1}{2}E(\Omega = 0). \tag{2.05}$$

Since for $\Omega\tau/2 = 1.4$ we have $\left[\frac{\sin(\Omega\tau/2)}{\Omega\tau/2}\right]^2 \approx 1/2$, the bandwidth of excitation can be defined as $\delta\omega_{bw} = 5.6/\tau$ or in Hertz $\delta\nu_{bw} = \frac{2.8}{\pi\tau}$.

If a train of pulses is applied, the mathematical expression for the spectral density $E(\Omega)$ of a train of $(2n+1)$ rectangular pulses with the distance T is given by, cf. Fig. 2.1,

$$E(\Omega) = \frac{\omega_{rf}^2\tau}{2\pi\gamma^2\mu_0}\left\{\frac{\sin(\Omega\tau/2)}{\Omega\tau/2}\right\}^2\left\{1 + 2\sum_{k=1}^{n}\cos(k\Omega T)\right\}^2. \tag{2.06}$$

2.2 *Resonance Offset*

The theoretical description of the (demodulated) FID is accomplished by choosing the carrier frequency of the pulse for the transformation into the interaction representation, cf. Eqs. (1.32) and (1.45). Then, the difference between the resonance frequency of the nuclear spin, ω_L, and the carrier frequency, ω, of an applied rf pulse (which is also the reference frequency for the demodulation of the NMR signal) is called *resonance offset*. Therefore, for the *detection* of the NMR signal, in the absence of pulsing, the effect of a resonance offset is simply the result of the demodulation of the signal with a frequency different from the

Larmor frequency, which corresponds to a frequency shift of the signal in the Fourier transform spectrum. For the *excitation*, however, a resonance offset causes the mean of the spectral density to be shifted with respect to the resonance frequency of the spin, and between pulses, the phase of the magnetization in the x, y-plane to be changed with time, and thus, the relative phase with respect to the next pulse becomes time dependent, as well.

For spin-1/2 nuclei and a negligible homogeneous interaction the influence of a resonance offset during pulsing is typically described by introducing an "effective" magnetic rf field, B_{eff}. In this case the operator $U_i(t, t_0)$ of Eq. (1.35) follows from Eqs. (1.52) and (1.54),

$$U_i(t, t_0) = \exp\{-i(\Delta\omega I_z + \omega_{rf} I_y)(t - t_0)\}, \tag{2.07}$$

which is equivalent to a rotation in the 3-dimensional space: Comparison with Eq. (1.09) shows that $n_y\Omega = -\omega_{rf}(t - t_0)$, and $n_z\Omega = -\Delta\omega(t - t_0)$. From the normalization of the axis of rotation, $\boldsymbol{n}$, we have

$$1 = \boldsymbol{n}^2 = (\omega_{rf}^2 + \Delta\omega^2)\frac{t - t_0}{\Omega^2}, \tag{2.08}$$

and

$$|\Omega| = |t - t_0|\sqrt{\omega_{rf}^2 + \Delta\omega^2}. \tag{2.09}$$

The angle between this axis of rotation, $\boldsymbol{n}$, and the laboratory z-axis is

$$\alpha = -\arcsin\left[\frac{\omega_{rf}}{\sqrt{\omega_{rf}^2 + \Delta\omega^2}}\right]. \tag{2.10}$$

For the angular frequency, ω_{eff}, of this rotation about the axis $\boldsymbol{n}$, we find

$$\omega_{eff} = \sqrt{\omega_{rf}^2 + \Delta\omega^2}. \tag{2.11}$$

This leads to the definition of the effective magnetic rf field, $B_{eff} = \omega_{eff}/\gamma$, where γ is the gyromagnetic ratio for the nucleus.

By means of the above shown equivalence, the effect of a resonance offset can be demonstrated by rotating the magnetization about an effective magnetic field, B_{eff}, which is inclined at an angle α compared to the laboratory z-axis ($\alpha = 90°$ for $\Delta\omega = 0$).

The picture of the effective magnetic field can also be applied if the resonance line is inhomogeneously broadened, but, for quadrupolar broadened lines it fails, since the frequency components of the NMR signal belong to different transitions in the spin system. Only if for a single crystal each transition can be excited separately, it would be possible to define effective magnetic rf fields for a single transition. However, for the more interesting case of partly selective excitation spin-flipping in one transition (due to the rf field) is closely related to the spin-flipping in another transition, so that the influence of the resonance offset during pulsing or between pulsing cannot be simplified. Therefore, we discuss the influence of the resonance offset together with the actual pulse experiment of the following chapters.

2.3 Nonselective Excitation

A "nonselective" excitation must be defined by two restrictions. First, in order to predominantly drive the evolution of the spin system by the interaction with the irradiated rf field, the pulse has to be strong compared with the internal interactions

$$\|\mathcal{H}_{rf}\| \gg \|\mathcal{H}_Q\|, \|\mathcal{H}_D\|, \|\mathcal{H}_{CSA}\|. \tag{2.12}$$

In this case the pulse is called a "hard" pulse. Second, the bandwidth, $\delta\nu_{bw}$, of excitation and the bandwidth of the probe circuit $\delta\nu_{probe}$ must be given by

$$\delta\nu_{bw}, \delta\nu_{probe} = \frac{\nu_L}{Q} \gg \frac{1}{h}\|\mathcal{H}_Q\|, \quad \frac{1}{h}\|\mathcal{H}_D\|, \quad \frac{1}{h}\|\mathcal{H}_{CSA}\|, \tag{2.13}$$

here, Q stands for the quality factor of the probe. If Eqs. (2.12) and (2.13) are fulfilled, the internal interactions within the spin system can be neglected during the time of excitation.

A pulse is called "partly selective" or "selective" if Eq. (2.13) does not hold. The pulse is termed "soft" if it does not meet Eq. (2.12).

Since for nonselective excitation we can neglect the influence of the quadrupole coupling during the action of a rf pulse, the Hamiltonian in the interaction representation describing the effect of a single y-pulse is time-independent, $\mathcal{H}_{rf,i} = \hbar\omega_{rf} I_y$, Eqs. (1.40) and (1.41) easily integrate to

$$U(t, t_0) = \exp\{-i\omega_{rf} I_y (t - t_0)\}. \tag{2.14}$$

If the spin system is in the thermal equilibrium before the pulse starts, at $t_0 = -\tau$, the corresponding density operator is $|\rho_0\rangle = |I_z\rangle$, Eq. (1.50). After a pulse with the duration τ, one gets at $t = 0$

$$|\rho_i(\tau)\rangle = \exp\{-i\omega_{rf} I_y \tau\}|I_z\rangle = \cos(\omega_{rf}\tau)|I_z\rangle + \sin(\omega_{rf}\tau)|I_x\rangle, \tag{2.15}$$

and, for the intensity of the FID, $G(t = 0)$, it is inferred from Eq. (1.51)

$$G(0) = \frac{\langle I_- | \rho_i(\tau)\rangle}{\langle I_x | I_x\rangle} = \sin(\omega_{rf}\tau). \tag{2.16}$$

After the pulse, $t \geqq 0$, the quadrupole interaction is assumed to govern the evolution of the density operator. If in the interaction representation the truncated quadrupole coupling is used $\mathcal{H}_Q^{(0)}$, Eq. (1.57), the evolution operator $U(t, t_0)$ is again obtained by integrating Eq. (1.34):

$$U(t, 0) = \exp\left\{-i\frac{\omega_Q'}{6}(3I_z^2 - I(I+1))t\right\}. \tag{2.17}$$

And with Eq. (2.17) and Eq. (2.15), $G(t)$ is given by

$$G(t) = \frac{1}{\langle I_x | I_x\rangle} \sum_{m=-I}^{I-1} 2W_m \langle m|\rho_i(\tau)|m+1\rangle \exp\{i\omega_Q'(m + 1/2)t\}, \tag{2.18}$$

where the trace is taken in the I_z representation, and

$$W_m = \tfrac{1}{2}\sqrt{I(I+1) - m(m+1)} = \tfrac{1}{2}\sqrt{(I-m)(I+m+1)}. \tag{2.19}$$

The free induction decay (FID) for quadrupole coupling, Eq. (2.18), consist of the $2I$ components (transitions) with amplitudes given by $W_m\langle m|\rho_i(\tau)|m+1\rangle$. In the notation of Eq. (2.18), the central transition corresponds to $m = -1/2$ since in this case the time-dependence in Eq. (2.18) vanishes. One also finds a different notation in the literature which follows from Eq. (2.18) by substitution of m by $m-1$. Both notations are essentially the same, the difference consists in whether one calculates the process $m \to m+1$ or $m+1 \to m$, which differ only by a phase factor, e.g., if the FID was a complex function, the sign of the imaginary part would be different for both processes.

The central transition, $m = -1/2$, is time-independent (to first-order), and has the intensity

$$G_{-1/2,1/2}(0) = \sin(\omega_{rf}\tau)\frac{2(W_{-1/2})^2}{\mathrm{tr}\{I_x^2\}} = \frac{3(2I+1)}{8I(I+1)}\sin(\omega_{rf}\tau), \tag{2.20}$$

where $\mathrm{tr}\{I_x^2\} = \tfrac{1}{3}I(I+1)(2I+1)$ has been used. Generally, the intensity of the transition, $m \leftrightarrow m+1$, is given by

$$G_{m,m+1}(0) = \sin(\omega_{rf}\tau)\frac{3\{I(I+1) - m(m+1)\}}{2I(I+1)(2I+1)}. \tag{2.21}$$

The dependence of the intensity of the FID, $G(0)$, is sinusoidal on ω_{rf} and τ for all transitions, as for spin $-1/2$ nuclei. The ratio of the intensities of two transitions is given by the square of the concerned matrix elements of I_x.

2.4 *Selective Excitation of a Single Transition*

For most non-cubic substances a nonselective excitation is not achievable. In such cases the exciting pulse is called a "soft" pulse, $\|\mathscr{H}_Q\| \gtrsim \|\mathscr{H}_{rf}\|$. For a single crystal with a sufficiently large quadrupole coupling it is always possible to tune to each transition, separately. But for powders, a selective pulse can only be applied to excite the central transition. Since the effect of the quadrupole interaction on the NMR spectrum depends on the orientation of the crystallite with respect to the static field, for powders, there will always be a number of satellite transitions within the spectral range of the central transition. For a selective excitation of the central transition, the fraction of excited satellites must be sufficiently small.

The excitation of a single transition for pure quadrupole coupling can be treated as a fictitious spin-1/2 system [1, 50] by choosing the appropriate 2 by 2 submatrix out of the corresponding $(2I+1)$ dimensional matrix and decompose it into spin-1/2 matrices, S_x, S_y, S_z. As an example, consider the $(m \leftrightarrow m+1)$

transition: for the submatrix out of I_z one has,

$$\begin{pmatrix} m+1 & 0 \\ 0 & m \end{pmatrix} = \frac{2m+1}{2}\begin{pmatrix} 1 & 0 \\ 0 & 1 \end{pmatrix} + \begin{pmatrix} 1/2 & 0 \\ 0 & -1/2 \end{pmatrix} = \frac{2m+1}{2}\mathbb{1} + S_z. \tag{2.22}$$

For the submatrix of I_z^2

$$\begin{pmatrix} (m+1)^2 & 0 \\ 0 & m^2 \end{pmatrix} = \frac{2m^2+2m+1}{2}\mathbb{1} + (2m+1)S_z, \tag{2.23}$$

and, for the y-component, I_y,

$$\begin{pmatrix} 0 & -\mathrm{i}W_{\mathrm{m}} \\ \mathrm{i}W_{\mathrm{m}} & 0 \end{pmatrix} = \frac{W_{\mathrm{m}}}{V_{-1/2}}\begin{pmatrix} 0 & -\mathrm{i}V_{-1/2} \\ \mathrm{i}V_{-1/2} & 0 \end{pmatrix} = \frac{W_{\mathrm{m}}}{V_{-1/2}}\cdot S_y = 2W_{\mathrm{m}}\cdot S_y, \tag{2.24}$$

where, W_{m} is defined by Eq. (2.19), and, $V_{-1/2} = W_{-1/2}(I=1/2) = \frac{1}{2}$.

The analog to Eq. (2.15) for selective excitation reads

$$\begin{aligned} |\sigma_{\mathrm{i}}(\tau)\rangle &= \exp\{-\mathrm{i}2W_{\mathrm{m}}\omega_{\mathrm{rf}}S_y\tau\}|S_z\rangle \\ &= \frac{2m+1}{2}\mathbb{1} + \cos(2W_{\mathrm{m}}\omega_{\mathrm{rf}}\tau)|S_z\rangle + \sin(2W_{\mathrm{m}}\omega_{\mathrm{rf}}\tau)|S_x\rangle. \end{aligned} \tag{2.25}$$

Finally, we get for the intensity of the NMR signal of the $m \leftrightarrow m+1$ transition after a selective pulse with a flip angle $\omega_{\mathrm{rf}}\tau$

$$\begin{aligned} G_{\mathrm{m,m+1}}(0) &= \frac{1}{\langle I_x | I_x \rangle} 2W_{\mathrm{m}}\langle -1/2|\sigma_{\mathrm{i}}(\tau)|1/2\rangle \sin(2W_{\mathrm{m}}\omega_{\mathrm{rf}}\tau) \\ &= \frac{3W_{\mathrm{m}}}{I(I+1)(2I+1)}\sin(2W_{\mathrm{m}}\omega_{\mathrm{rf}}\tau), \end{aligned} \tag{2.26}$$

because the contributions arising from the first and second term on the right hand side of Eq. (2.25) vanish. Comparison with Eq. (2.21) reveals that the maximal observed intensity is reduced by $2W_{\mathrm{m}}$, but, the nutation frequency is enhanced, as

$$\tilde{\omega}_{rf} = 2W_{\mathrm{m}}\omega_{\mathrm{rf}}. \tag{2.27}$$

For the central transition, $m = -1/2$, this yields

$$\tilde{\omega}_{\mathrm{rf}} = (I+1/2)\omega_{\mathrm{rf}}. \tag{2.28}$$

Expressions like "solid" or "liquid" π-pulse for $\tilde{\omega}_{\mathrm{rf}}\tau = \pi$ and $\omega_{\mathrm{rf}}\tau = \pi$, respectively, will not be used in this review because they are often misused in literature.

In the photon language, the understanding of nonselective and selective excitation is the following: The spin-flipping within $2I+1$ energy levels due to the excitation corresponds to absorption and stimulated emission of photons. The number of photons in the spectral range of excitation is proportional to the spectral energy density $E(\Omega)$, Eq. (2.04). If the width of the frequency distribution of the photons, $E(\Omega)$, is large compared to the spectral range of transitions for the spin system the excitation is nonselective. At the time after a π-pulse the population difference is inverted, or, after a $\pi/2$-pulse the population difference is zero. In the high temperature approximation, the population number N_m of the Zeeman level E_m can be written, as $N_m = 2m$. Then, e.g., for $I = 5/2$ the population of the $2I + 1 = 6$ levels is $-5, -3, -1, 1, 3, 5$. The conservation of energy, together with the selection rule $\Delta m = \pm 1$, shows that after a $\pi/2$-pulse in each of the $2I = 5$ transitions, $m \rightarrow m + 1$, the net number of absorbed photons is 5, (5 + 3), (5 + 3 + 1), (5 + 3), 5 for $m = -5/2, -3/2, -1/2, 1/2, 3/2$. These are the relative intensities of the 5 transitions for nonselective excitation. For a selective excitation of the central transition the net number of absorptions after a $\pi/2$-pulse is 1, as for any other single transition. In order to compare the number of absorptions with those for nonselective excitation we have to recall that the observed intensity is proportional to the number of absorbed photons per unit time and proportional to the number of photons incident with the appropriate frequency [51]. Suppose ω_{rf} remains constant, then, the difference in time for reaching zero population difference for selective and nonselective excitation can be deduced from Eq. (2.04). Since the number of incident photons, $E(\Omega \approx 0) \propto \tau$, increases with time, as well, the total number of absorptions being quadratic in τ, we find for the pulse durations τ_n and τ_s, necessary to cancel the population differences for nonselective and selective excitation, respectively, the ratio $(\tau_n/\tau_s)^2 = 9/1$ for $I = 5/2$. If τ remained constant for both kinds of excitations and ω_{rf} was changed, in order to reach zero population difference, one would get the same result since $E(\Omega) \propto \omega_{rf}^2$. Thus, for comparison of the maximum intensities for the central transitions, for both, nonselective and selective excitation, I_n and I_s, one finds in accordance with Eqs. (2.20) and (2.26) for $I = 5/2$ that $I_n/I_s = 9\tau_s/\tau_n = 3$.

2.5 *Partly Selective Excitation of More Than One Transition*

In practice, a pure selective excitation of a single transition cannot be achieved for powder samples. Some nuclei experience an electric field gradient tensor with z-axis placed not far from the magic angle refered to the external magnetic field, that means $\omega_Q' \approx 0$ in Eq. (1.22). Thus, some satellite transitions will occur even if selective excitation is achieved. On the other hand one faces the problem in approaches which make use of the partly selective excitation for the determination of quadrupole parameters, e.g. the nutation technique. If the spectral width of the central transition is small compared with ω_{rf}, second-order quadrupole effects can be neglected. Then, the problem is to analyze the following equation

$$|\rho(\tau)\rangle = \exp\{i[\omega_{rf} I_y + \tfrac{1}{6}\omega_Q'(3I_z^2 + I(I+1))]\tau\} |I_z\rangle, \tag{2.29}$$

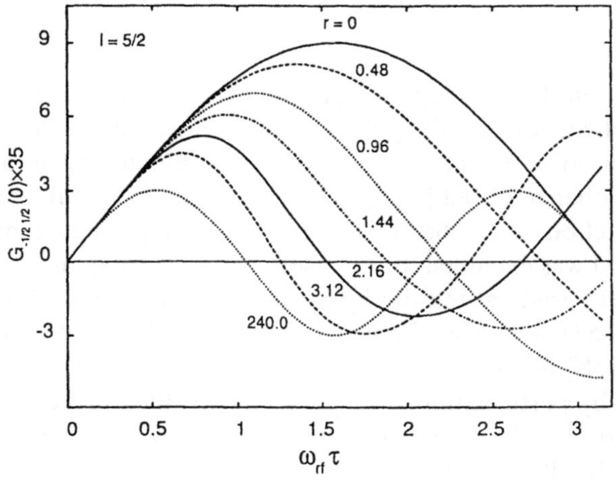

Fig. 2.2. Dependence of the intensity of the central line on $\omega_{rf}\tau$ and $r = \omega_Q'/\omega_{rf}$

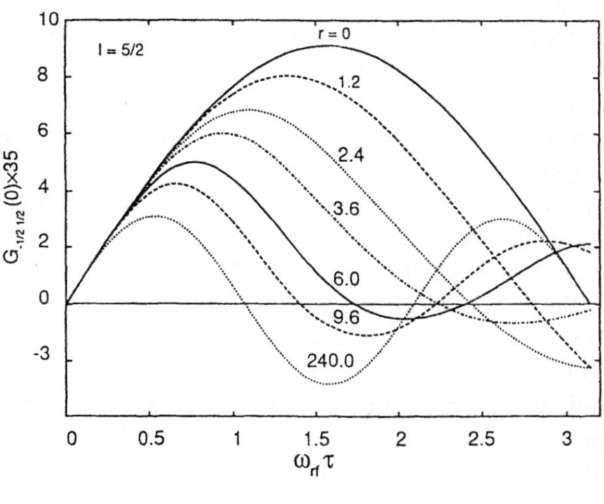

Fig. 2.3. Dependence of the intensity of the central line on $\omega_{rf}\tau$ and the powder average of ω_Q'. $r = \omega_Q/\omega_{rf}$

in which the influence of the quadrupole Hamiltonian (in first-order) during the pulse is being considered. Since an analytical solution of Eq. (2.29) is rather complicated, we show below numerical results for the dependence of the intensity of the central line on ω_{rf}/ω_Q', Fig. 2.2, cf. [52, 53]. The results are presented in the powder average, as well, Fig. 2.3. In the latter Figure one realizes the dependence for nonselective ($\omega_Q \to 0$) and selective ($\omega_Q \to \infty$) excitation. Nielsen et al. [37] considered the excitation of $I = 3/2$ and $I = 5/2$ nuclei in the case of MAS. They present theoretical curves similar to those in Fig. 2.2 for $I = 3/2$ but as a function of the spinning speed. Since under MAS the (orientation dependent) EFG is averaged over the duration of the pulse, the quadrupole interaction appears to be smaller, depending on spinning speed and pulse duration.

3 Lineshapes

A wealth of literature exists on the lineshape of half-integer quadrupole nuclei. For a symmetric field gradient, the quadrupolar lineshape was calculated in closed form for first-order and second-order pattern by Bloembergen as presented in the review of Cohen and Reif [5]. Spectra describing the combined effect of quadrupole interactions and chemical shift were first presented by Jones et al. [54]. The method was extended to the case of non-axially symmetry of the chemical shift and field gradient tensors by Stauss [55], by Narita et al. [56], and by Baugher et al. [57]. A review of powder spectra in NMR and EPR was given by Taylor et al. [58] in 1975. For stronger quadrupole interactions, which cannot be described by perturbation theory, the eigenvalue problem of the full Hamiltonian has to be solved, as was demonstrated by Abart et al. [59, 60]. Barnes et al. [61] briefly reviewed the work in this field up to 1988 which had been carried out under the condition that the quadrupole and the chemical shift tensors are coincident. Several examples of low-symmetry sites were studied which do not meet this condition and simulations were made for non-coincident tensors by Chu and Gerstein [62], by Cheng et al. [63], and Power et al. [64] describing the lineshapes of ^{85,87}Rb in rubidium salts, and ^{133}Cs in cesium chromate, respectively. A corresponding ^{51}V MAS study was performed by Skibsted et al. [65].

A new procedure for the powder average has been introduced by Alderman et al. [66]. This approach provides a very fast simulation of powder patterns since, first, the intensity for a special orientation is averaged over the next nearest orientations, second, the time-consuming calculation of trigonometric functions could be replaced by simple division. Zheng et al. [43] extended this approach and the one introduced by Sethi et al. [67], for the computation of sideband intensities to a simple formalism for computing NMR spectra of the central transition of quadrupole nuclei with half-integer spin in powdered solids spinning at any angle and at any speed. Some aspects of the lineshape of the central transition of spinning samples will be discussed in Sect. 6.

The lineshape of the NMR signal after a single pulse is given by the Fourier transform of its free induction decay. With Eqs. (1.51) and (1.33) we have for the free induction decay

$$G(t) = \frac{\langle I_- | \rho_i(t) \rangle}{\langle I_x | I_x \rangle} = \frac{\langle I_- | U(t, t_0) \rho(t_0) U(t, t_0)^{-1} \rangle}{\langle I_x | I_x \rangle}, \tag{3.01}$$

where $\rho(t_0 = 0)$ represents the density operator after the pulse, $t_0 = 0$. As a good approximation one can consider only operators in $U(t)$ which are secular with respect to the Zeeman interaction. Then we find with Eq. (3.01)

$$G(t) = \frac{1}{\mathrm{tr}\{I_x^2\}} \sum_{m=-I}^{I-1} 2W_m \rho_{m,m+1} U_{m,m} U^{-1}_{m+1,m+1}. \tag{3.02}$$

Since $U = \exp\left\{ -\frac{i}{\hbar} \mathscr{H} t \right\}$ is diagonal in the set $|m\rangle$, the Fourier transform (defined by $\int_{-\infty}^{+\infty} G(t)\exp\{-\mathrm{i}\Omega t\}\,\mathrm{d}t$) of Eq. (3.02) yields for the lineshape, $F(\Omega)$, of the observed signal

$$F(\Omega)=\frac{1}{\mathrm{tr}\{I_x^2\}}\sum_{m=-I}^{+I-1} 2W_m\rho_{m,m+1}\,\delta(\Omega-\{\mathscr{H}_{m+1,m+1}-\mathscr{H}_{mm}\}/\hbar). \qquad (3.03)$$

The different interactions, $\mathscr{H}$ (in the interactions representation, cf. Eqs. (1.52) ff.), contain the frequency distribution. The matrix elements of the density operator, $\rho_{m,m+1}$, contain the information about the excitation. Since $G(t)$ represents the demodulated NMR signal, it is $\Omega=\omega-\omega_L$, and $\{\mathscr{H}_{m+1,m+1}-\mathscr{H}_{mm}\}/\hbar=\omega_{m+1,m}-\omega_L$, where ω_L is the Larmor frequency.

3.1 First-Order Quadrupole Shift

For the first-order quadrupole interaction we consider $\mathscr{H}=\bar{\mathscr{H}}_Q^{(0)}$ as given by Eq. (1.57), assuming the resonance offset to be zero. Then, the first-order frequency shift becomes

$$\begin{aligned}\omega_{m,m+1}-\omega_L&=\omega_Q\left\{\frac{3\cos^2\beta-1}{2}+\frac{\eta}{2}\sin^2\beta\cos 2\alpha\right\}\left(m+\frac{1}{2}\right)\\&=\omega_Q'\left(m+\frac{1}{2}\right).\end{aligned} \qquad (3.04)$$

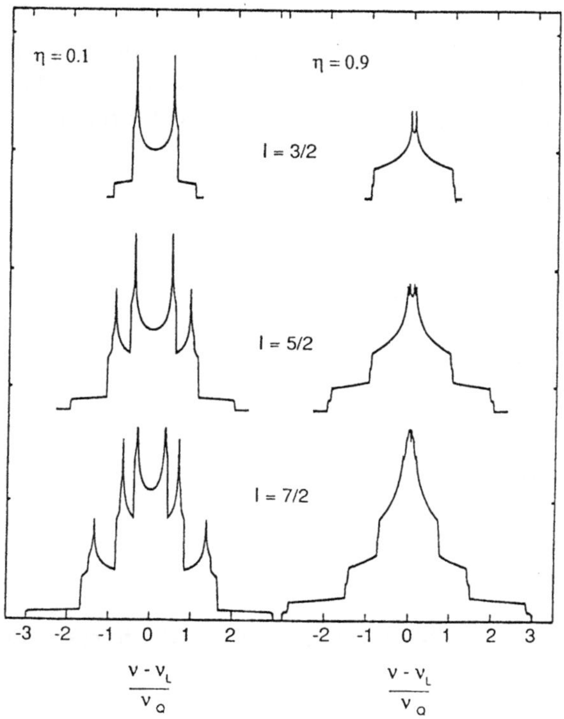

Fig. 3.1. Examples for the lineshape of first-order quadrupole spectra for nonselective excitation. The central line is omitted

It should be mentioned again that in this notation the transition $m \rightarrow m+1$ is considered so that the central transition corresponds to $m = -1/2$, cf. remarks below Eq. (2.19).

In order to calculate the lineshape one has to make assumptions about the excitation, cf. Sect. 2, and the distribution of the Euler angles α, β. If nonselective excitation applies, e.g., one has for a $\pi/2$-pulse in y-direction $\rho = -I_x$. For partly selective excitation $\rho_{m,m+1}$ has to be determined numerically, as is shown by Eq. (2.29). Fig. 3.1 gives some examples for the lineshape of first-order quadrupole spectra for nonselective excitation, $F_{ns}(\omega - \omega_L)$, and samples of random powder. The sharp line at the Larmor frequency due to the central transition is omitted in this figure.

3.2 *Second-Order Quadrupole Shift*

For finding the second-order quadrupole shift, in Eq. (3.03) we have to consider $H = \bar{\mathscr{H}}_Q^{(1)}$, with $\bar{\mathscr{H}}_Q^{(1)}$ given by Eq. (1.58). Calculation of $\{\mathscr{H}_{m+1,m+1} - \mathscr{H}_{mm}\}/\hbar$, cf. Eq. (3.03), yields (carrier frequency of the pulse, ω, equals the Larmor frequency ω_L)

$$\omega_{m,m+1} - \omega_L = -\frac{\omega_Q^2}{9\omega_L}\left\{\left[12m(m+1) - 2I(I+1) + \frac{9}{2}\right]V_1 V_{-1} + \left[3m(m+1) - I(I+1) + \frac{3}{2}\right]V_2 V_{-2}\right\}. \tag{3.05}$$

Taulelle [68] expressed the products $V_2 V_{-2}$ and $V_1 V_{-1}$ in terms of the principle axis components. For the central transition, $m = -1/2$, both squared brackets in Eq. (3.05) consist of multiples of $I(I+1) - 3/4$, and one finds for the matrix elements

$$\omega_{-1/2,1/2} - \omega_L = -\frac{\omega_Q^2}{6\omega_L}\left\{I(I+1) - \frac{3}{4}\right\}\{A(\alpha,\eta)\cos^4\beta + B(\alpha,\eta)\cos^2\beta + C(\alpha,\eta)\}, \tag{3.06}$$

where

$$A = -\frac{27}{8} - \frac{9}{4}\eta\cos 2\alpha - \frac{3}{8}\eta^2\cos^2 2\alpha,$$

$$B = +\frac{15}{4} - \frac{1}{2}\eta^2 + 2\eta\cos 2\alpha + \frac{3}{4}\eta^2\cos^2 2\alpha,$$

$$C = -\frac{3}{8} + \frac{1}{3}\eta^2 + \frac{1}{4}\eta\cos 2\alpha - \frac{3}{8}\eta^2\cos^2 2\alpha. \tag{3.07}$$

Equations (3.06–3.07) were first used by Narita et al. [56] in order to calculate the second-order powder pattern of the central transition for a static sample.

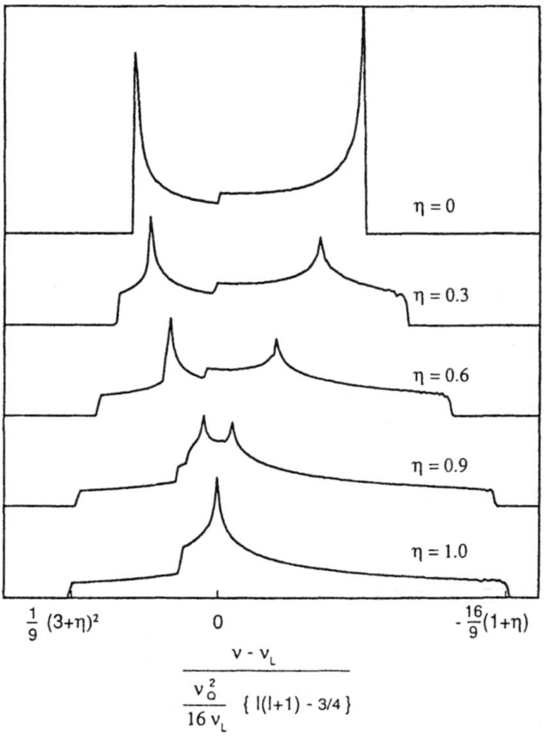

Fig. 3.2. Examples for the lineshape of second-order quadrupole spectra of the central transition

In Fig. 3.2 we show some examples of second-order lineshapes for random powder.

If one can assume that the spinning speed in a sample spinning experiment is larger than the second-order quadrupole shift, the irreducible tensor components V_2V_{-2} and V_1V_{-1}, contained in $\bar{\mathscr{H}}_{\mathrm{Q}}^{(1)}$ of Eq. (1.58), can be transformed to the principle axis system of the EFG tensor via the system of the spinning rotor and the time-average of these terms can be used in order to express the influence of a fast sample rotation. As was shown by Müller [69], for the case of spinning at the magic angle, these result can be expressed by introducing the coefficients A^{MAS}, B^{MAS} and C^{MAS} instead of A, B, C in Eqs. (3.06–3.07). These new coefficients are

$$A^{\mathrm{MAS}} = +\frac{21}{16} - \frac{7}{8}\eta\cos 2\alpha + \frac{7}{48}\eta^2\cos^2 2\alpha,$$

$$B^{\mathrm{MAS}} = -\frac{9}{8} + \frac{1}{12}\eta^2 + \eta\cos 2\alpha - \frac{7}{24}\eta^2\cos^2 2\alpha,$$

$$C^{\mathrm{MAS}} = +\frac{15}{16} - \frac{1}{8}\eta\cos 2\alpha + \frac{7}{48}\eta^2\cos^2 2\alpha. \qquad (3.08)$$

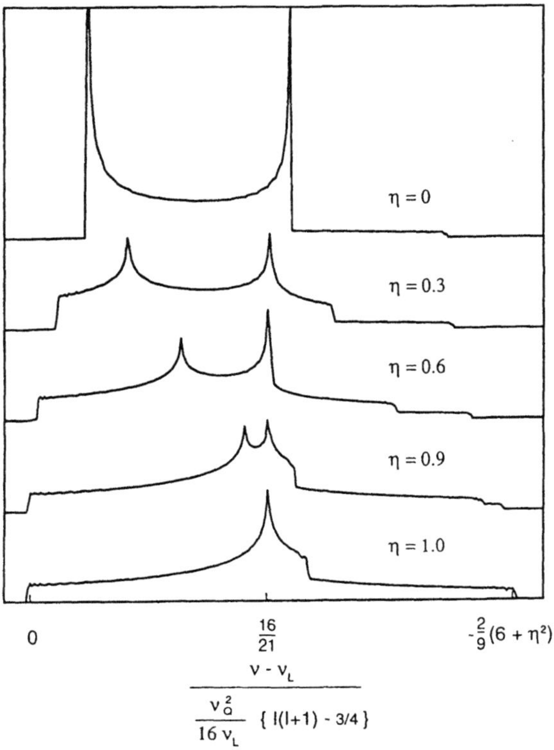

Fig. 3.3. Examples for the lineshape of second-order quadrupole spectra of the central transition for sample rotation about the magic angle

Numerically calculated lineshapes for sample rotation about the magic angle, MAS, are presented in Fig. 3.3. Significant features of these spectra are given in terms of the following expressions [69]:

$$\omega_1 - \omega_L = -\frac{2}{9}(6 + \eta^2)A \qquad \text{minimum frequency,}$$

$$\omega_2 - \omega_L = -\frac{1}{18}(15 + 6\eta + 7\eta^2)A \qquad \text{first shoulder,}$$

$$\omega_3 - \omega_L = -\frac{1}{18}(15 - 6\eta + 7\eta^2)A \qquad \text{singularity for } \eta \leqq \frac{3}{7} \text{ or shoulder for } \eta > \frac{3}{7},$$

$$\omega_4 - \omega_L = -\frac{16}{21}A \qquad \text{singularity for } \eta \geqq \frac{3}{7},$$

$$\omega_5 - \omega_L = -\frac{4}{21}(1 + \eta)^2 A \qquad \text{second singularity,}$$

$$\omega_6 - \omega_L = -\frac{4}{21}(1 - \eta)^2 A \qquad \text{maximum frequency,} \qquad (3.09)$$

where

$$A = \frac{\omega_Q^2}{16\omega_L}\left[I(I+1) - \frac{3}{4}\right] \tag{3.10}$$

is the convenient measure for the width of the second-order broadened spectrum, which was introduced by Cohen and Reif [5]. It should not be confused with function A in Eqs. (3.06–3.07). On this scale, the spectral width for the MAS spectra is $72A/63$ and $98A/63$ for $\eta = 0$ and 1, respectively, in comparison to $25A/9$ and $40A/9$ for the static spectra, $\eta = 0$ and 1, respectively.

In order to deduce quadrupole parameters from experimentally obtained MAS spectra, it is usual to fit the experimental spectrum with calculated spectra using a utility program of the spectrometer software. Engelhardt and Koller [70] proposed a simple procedure without spectral simulation.

If the spinning frequency is not large compared with the second-order frequency shift, the average Hamiltonian according to Eq. (1.39) has to be calculated. Fig. 3.4 shows spectra calculated by Zheng et al. [43].

The Eqs. (3.06–3.07) allow an analytical determination of the center of gravity for powder patterns, as well as for the corresponding second moments of the

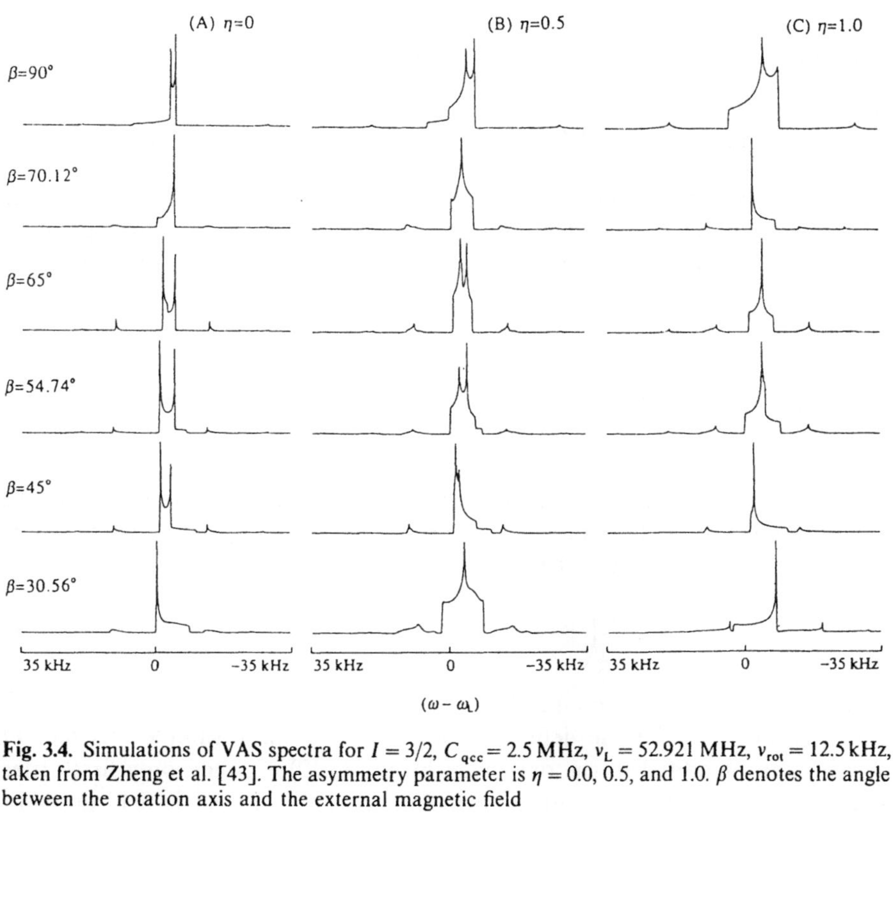

Fig. 3.4. Simulations of VAS spectra for $I = 3/2$, $C_{qcc} = 2.5$ MHz, $\nu_L = 52.921$ MHz, $\nu_{rot} = 12.5$ kHz, taken from Zheng et al. [43]. The asymmetry parameter is $\eta = 0.0$, 0.5, and 1.0. β denotes the angle between the rotation axis and the external magnetic field

line shapes [71]. A single expression, describing the shift of the center of gravity of both, the static and MAS case, is obtained in agreement with earlier results [54, 72]

$$\Delta\omega_{cg} = \omega_{cg} - \omega_L = -\frac{\omega_Q^2}{30\omega_L}[I(I+1) - 3/4]\left(1 + \frac{\eta^2}{3}\right). \tag{3.11}$$

The corresponding value in the A-scale is

$$\Delta\omega_{cg} = \omega_{cg} - \omega_L = -\frac{8A}{15}\left(1 + \frac{\eta^2}{3}\right). \tag{3.12}$$

For the second moments, M_2, of the line shape of the central transition with respect to the center of gravity the following expressions can be derived: For the static sample

$$M_2 = \Delta\omega_{cg}^2 \cdot \frac{23}{7}, \tag{3.13}$$

for the case of MAS, neglecting spinning sidebands,

$$M_2^{\mathrm{MAS}} = \Delta\omega_{cg}^2 \cdot \frac{1}{4}. \tag{3.14}$$

From Eqs. (3.13) and (3.14) it is found that

$$\sqrt{\frac{M_2}{M_2^{\mathrm{MAS}}}} = \sqrt{\frac{92}{7}} \approx 3.6, \tag{3.15}$$

which describes the reduction of the central transition linewidth caused by the second-order quadrupole interaction by means of MAS [71]. Corresponding values for other angles than the magic one are given in Sect. 6, Fig. 6.1.

3.3 Contributions from Dipole Interactions and Chemical Shift Anisotropy

Dipolar interactions and the anisotropy of chemical shift also contribute to the second moment of a static powder spectrum. The second moment with respect to the isotropic value of the chemical shift is

$$M_2^{\mathrm{CSA}} = \frac{4}{9}\frac{(\Delta\sigma\omega_{\mathrm{L}})^2}{5}\left(1 + \frac{\eta^2}{3}\right), \tag{3.16}$$

where $\Delta\sigma = \sigma_{ZZ} - (\sigma_{XX} + \sigma_{YY})/2$ denotes the anisotropy of the chemical shift tensor, cf. Sect. 1.5 for the definition of η. For the static sample this second moment can be added to the one given by Eq. (3.12) if either: only first-order effects must be considered and the principal axes of the electric field gradient tensor and the chemical shift tensor are coincident, or, the chemical shift anisotropy is so small that off-diagonal terms can be neglected, whereas the central line is broadened by second-order quadrupole interaction [57]. For the general case see the literature in the first paragraph of Sect. 3 [62–65].

If the dipolar interaction of a spin pair with the distance r_{ik} is large compared with the quadrupole interaction, the second moment of the corresponding lineshape is in the powder average given in Ref. [1] for like spin

$$M_2^{II} = \frac{9}{5}\left(\frac{\mu_0}{4\pi}\gamma_I^2\hbar\right)^2 F(I)\frac{1}{r_{ik}^6} = \frac{9}{5}\left(\frac{\mu_0}{4\pi}\gamma_I^2\hbar\right)^2 \frac{I(I+1)}{3}\frac{1}{r_{ik}^6}, \tag{3.17}$$

and for unlike spin

$$M_2^{IS} = \frac{4}{5}\left(\frac{\mu_0}{4\pi}\gamma_I\gamma_S\hbar\right)^2 \frac{S(S+1)}{3}\frac{1}{r_{ik}^6}, \tag{3.18}$$

where I, S denote the spins of the resonant and non-resonant nuclei, respectively.

As quadrupole coupling becomes stronger, the spin-flipping between different transitions is prohibited, and the second moment changes. For the central transition of a quadrupole nucleus this effect has been treated theoretically by Kambe and Ollom [73] and by Mansfield [74]. The heteronuclear contribution to the second moment remains the same, if a quadrupole interaction is present [1]. The influence of the quadrupole interaction on the second moment of the homonuclear dipole interaction can be expressed using Van Vleck's formula for identical spins as given in Eq. (3.17) by introducing different functions $F(I)$. If the notation of Abragam is used (Ref. [1], p. 130), it is $F(I) = I(I+1)/3 = F_D$ for pure dipolar interaction in the absence of quadrupole interaction. If all spins are subjected to the same quadrupole coupling (same magnitude and orientation of the electric field gradient) the spins have been called "like spins" and the corresponding factor, $F(I) = F_L$, is given by

$$F_L = \frac{4I(I+1)}{27} + \frac{[2I^2(I+1)^2 + 3I(I+1) + 13/8]}{18(2I+1)}. \tag{3.19}$$

However, if the quadrupole coupling of the two spins is different, but the central transition frequencies are still the same, the spins may be called "semi-like spins" and the corresponding factor, $F(I) = F_{SL}$, is given by

$$F_{SL} = \frac{4I(I+1)}{27} + \frac{(2I+1)}{18} + \frac{(2I+1)^3}{288}. \tag{3.20}$$

With $F_{UL} = 4I(I+1)/27$ (hypothetical unlike spins which can be described by one value of γ), comparison of the factors shows $F_D > F_L > F_{SL} > F_{UL}$, and a difference of less than 20% between F_L and F_{SL}. A similar change of the second moment of the satellite transitions can be calculated [75].

4 Nutation Spectroscopy

Torrey [76] studied, as early as 1949, the nutation of the resultant nuclear magnetic moment vector by applying radio frequency pulses with a carrier frequency close to the resonance frequency of the spins. During the last decade nutation spectroscopy has been the subject of much research. A few papers are concerned with dipole coupled systems [77, 78], imaging [79] or ^{2}H NMR [80] but the majority of nutation studies is dedicated to half-integer quadrupolar nuclei following the work of Samoson and Lippmaa who introduced two-dimensional nutation NMR [81, 82]. In NQR, two-dimensional nutation spectroscopy is also used [83, 84].

The simple 2D-experiment is divided into the evolution period t_1 during which a strong rf field is irradiated and, a detection period t_2 which is a FID. In the rotating frame, the spins *nutate* (precess) around the strong rf field with specific nutation frequencies ω_1, here the subscript 1 denotes the frequency axis ω_1 in the 2D-spectrum corresponding to the Fourier transform with respect to t_1 and should not be confused with $\omega_{rf} = \gamma B_{rf}$ which is a constant for the 2D experiment.

For quadrupole nuclei the nutation frequencies ω_1 depend on the strength of the quadrupole interaction. If $\omega_Q \ll \omega_{rf}$, then the transverse magnetization responds to the rf pulse like spin-1/2 nuclei, thus $\omega_1 = \omega_{rf}$. If $\omega_Q \gg 10\omega_{rf}$ the central transition can be treated as a two-level system and one nutation frequency is expected, but, as described in Sect. 2, Eq. (2.28), this frequency is increased by a factor $I + 1/2$, i.e. $\omega_1 = (I + 1/2)\omega_{rf}$. For the intermediate case, $0.1 < \omega_Q/\omega_{rf} < 100$, the nutation spectra are complicated because of the influence of the partly excited outer transitions.

It should be noted here that a variety of different definitions of the quadrupole frequency exists in the literature. We use in agreement with Abragam [1] and Cohen and Reif [5] the definition

$$\frac{\omega_Q}{2\pi} = \nu_Q = \frac{3C_{qcc}}{2I(2I-1)} \tag{4.01}$$

with

$$C_{qcc} = \frac{e^2qQ}{h}$$

and

$$\omega'_Q = \omega_Q \cdot \tfrac{1}{2}(3\cos^2\beta - 1 + \eta\sin^2\beta\cos 2\alpha),$$

as the angular dependent quadrupole frequency, cf. Sect. 1, Eqs. (1.16) and (1.22). With these definitions ω_Q corresponds to the maximum first-order frequency shift of the $\pm 3/2 \leftrightarrow \pm 1/2$ transitions for a powder pattern which is the same as the frequency distance between the singularities in a powder pattern related to the $\pm 3/2 \leftrightarrow \pm 1/2$ transitions for $\eta = 0$. In order to review the nutation literature the various definitions used by other authors for the quadrupole frequency or the angular dependent quadrupole frequency must be clarified. If we denote with ω_Q^* those differently defined quadrupole frequencies it is: $\omega'_Q = 6\omega_Q^*$ or $\omega_Q = 6\omega_Q^{max}$ used by Samoson et al. [82] and Nielsen et al. [36]; $\omega'_Q = 12\omega_Q^*$ used by Veeman and coworkers [85–87]; $\omega_Q = 2\omega_Q^*$ used by Man [88]. These values must be taken into account for the comparison of the numerical results.

In the photon language, which we used in Sect. 2.4, the nutation experiment can be understood as follows: The Zeeman energy levels of half-integer spin nuclei are split due to quadrupole interaction resulting in $2I$ transitions, symmetric about the central one in first-order perturbation theory. By switching in a rf pulse photons are supplied, necessary for induced transitions between the $2I + 1$ energy levels. But, the spectral range of the available photons is inversely proportional to the pulse duration τ. Increasing τ means increasing the number of photons closer to the carrier frequency of the pulse, which should match the Larmor frequency. Although we are observing only the central transition, spins from the outer levels will pass the central ones contributing to the intensity of the central transition. But a sufficiently large number of photons must be available over the whole spectral range of resonances, which is about $(2I - 1)\omega_Q$, that means, a sufficient excitation should occur within the time τ, which is given by, cf. Sect. 2.1, $\delta\omega_{bw} = (2I - 1)\omega_Q = 5.6/\tau$. If we set the minimum flip angle for sufficient sensitivity to $\omega_{rf}\tau = \pi/25$, the rf amplitude should meet the condition

$$\frac{\omega_{rf}}{\omega_Q} \gtrsim \frac{(2I - 1)\pi}{140}. \tag{4.02}$$

Nutation spectra can be calculated using numerical procedures or by means of an analytical description which is the more complicated the larger the spin quantum number. The powder average produces quite complicated nutation spectra and it is necessary, in order to deduce quadrupolar parameters from the spectra, to compare the experimental spectrum with a set of calculated spectra.

4.1 Lineshape of Nutation Spectra

An analytical approach was found for $I = 3/2$ [89, 90] and for $I = 5/2$ [91] yielding $(I + 1/2)^2$ different nutation frequencies of the central transition. Each frequency gives rise to a powder pattern, and $2I$ of the $(I + 1/2)^2$ frequencies are dominant. The weaker components of the nutation frequencies of the central transition can be explained as multiple-quantum coherences in a 90° tilted rotating frame [92]. Since a 90° tilt about the y-axis of the rotated frame is performed by a transformation of the operators according to

$$\tilde{\mathscr{H}} = \exp\left\{i\frac{\pi}{2}I_y\right\}\mathscr{H}\exp\left\{-i\frac{\pi}{2}I_y\right\} \quad \text{and} \quad |\tilde{m}\rangle = \exp\left\{i\frac{\pi}{2}I_y\right\}|m\rangle, \tag{4.03}$$

the transformed first-order quadrupole Hamiltonian is given by [92]

$$\tilde{\mathscr{H}}_{\mathrm{Q}} = -\omega_{\mathrm{rf}}\tilde{I}_{\mathrm{z}} - \frac{\omega_{\mathrm{Q}}'}{12}\left\{3\tilde{I}_{\mathrm{z}}^2 - \tilde{I}^2 - \frac{3}{2}(\tilde{I}_{+}^2 - \tilde{I}_{-}^2)\right\}. \tag{4.04}$$

The comparison of Eq. (4.04) with Eq. (1.23) indicates that the secular part of the quadrupole interaction is reduced by a factor of 1/2. In the new frame of reference the elements of the density matrix at a time t_1, immediately after the pulse, can be approximated for not too long a t_1 if the rf power meets Eq. (4.02) by [92]

$$\tilde{\rho}_{\mathrm{m,m+1}} \propto \sqrt{I(I+1) - m(m+1)}\exp\left\{\mathrm{i}\left(\omega_{\mathrm{rf}} + \frac{\omega_{\mathrm{Q}}'}{2}(m+1/2)\right)t_1\right\}. \tag{4.05}$$

This clearly shows the modulation of the rotating frame central transition magnetization components. Since $|\tilde{m}\rangle$ is a sum of all $2I$ eigenfunctions $|m\rangle$ there will be $2I$ components, $m = I-1, \ldots, -I$, the amplitude of which depends on the Wigner rotation matrices $\langle m|\exp\{\mathrm{i}\pi I_y/2\}|m+1\rangle$.

In Figs. 4.1 and 4.2 we present some calculated lineshapes of nutation spectra for $I = 3/2$ and 5/2, respectively, obtained by numerical analysis according to Eq. (2.29), in the powder average, with a subsequent Fourier transform with respect to the pulse duration. For higher spin quantum numbers cf. Kentgens et al. [85].

Beyond the lineshape of the nutation spectra the "center of gravity" can be used for quantitative analysis [92], but the spectra are very broad and the error is quite large. It should be remarked, that the center of gravity of the nutation spectrum is a constant, which can be seen by taking the first derivative with

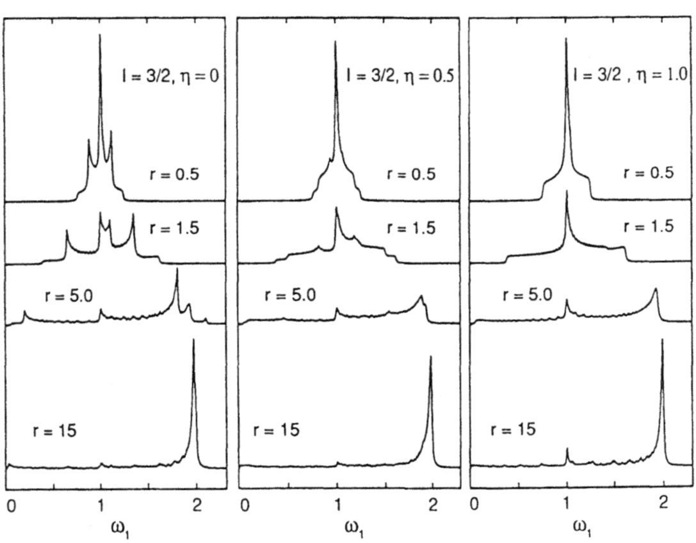

Fig. 4.1. Some calculated lineshapes of nutation spectra for $I = 3/2$ obtained by numerical analysis according to Eq. (2.29), in the powder average, and performing a subsequent Fourier transform with respect to the pulse duration

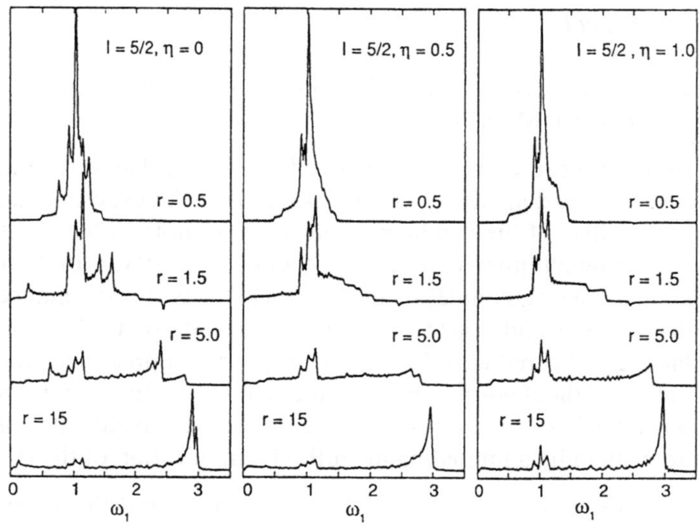

Fig. 4.2. Some calculated lineshapes of nutation spectra for 5/2, obtained by numerical analysis according to Eq. (2.29), in the powder average, and performing a subsequent Fourier transform with respect to the pulse duration

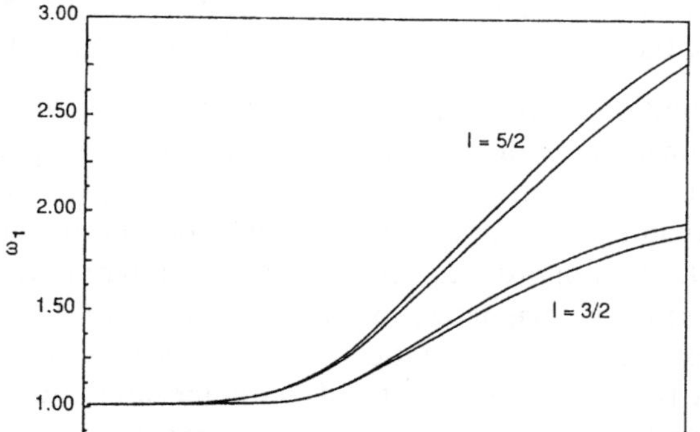

Fig. 4.3. Dependence of the center of gravity of the nutation spectra on the ratio $r = \omega_Q/\omega_{rf}$ for $I = 3/2$ and 5/2 and two values of η. The upper and lower curves correspond to $\eta = 0$ and $\eta = 1$, respectively

respect to the pulse length in Eqs. (2.20) and (2.26) for nonselective and selective excitation, respectively. But, by normalizing the calculated lineshapes of the nutation spectra, the "center of gravity" becomes a function of the excitation, so, the change in the "center of gravity" is truly a change in intensity. This can also be seen from Fig. 2.3: the gradient at $\omega_{rf}\tau = 0$ ($t_1 = 0$) with respect to τ is not a function of ω_Q/ω_{rf} what makes intensity measurements for short pulses independent on the excitation, but, the intensity of the oscillation changes with ω_Q/ω_{rf}. Figure 4.3 shows the dependence of the "center of gravity" on the ratio ω_Q/ω_{rf} for $I = 3/2$ and 5/2.

4.2 Experimental Aspects

The nutation spectrum is very sensitive to the following influences: Inhomogeneity of the rf field, resonance offset, MAS, T_1.

Inhomogeneity of the rf field causes a symmetric line broadening. The spread in the ratio ω_Q/ω_{rf} causes a spread in ω_1/ω_{rf}, cf. Figs. 4.1 and 4.2. An experimental approach to reduce the effect of the rf inhomogeneity is the subtraction of the signal for very large t_1 from all other signals [86]. Nielsen et al. [93] found for MAS nutation experiments a ring sample geometry to yield the best performance with respect to both sensitivity and rf field homogeneity. In addition to the local inhomogeneity which can be reduced by diminishing the sample size, the deviation of the envelope of the rf pulse from a rectangular shape causes artifacts in the nutation spectra. Such a deviation is usually caused by the bandwidth of the probe circuit rather than by a limited bandwidth of amplification of the rf.

The resonance offset changes the symmetry of the Hamiltonian and the signal becomes phase modulated which causes dispersive contributions to the spectrum. This effect can easily distort powder pattern [85]. Since the effect of an offset increases with increasing t_1 an artificial line appears at $\omega_1 = 0$. When adjusting the carrier frequency of the rf pulse to the Larmor frequency ω_L it must be taken into account that the center of gravity ω_{cg} of the second-order influenced central line is shifted with respect to ω_L by the quadrupole shift, cf. Sect. 6. For MAS spectra ω_L may even be outside the spectral range of the central line, cf. Sect. 3.

MAS enhances the spectral resolution for the single pulse experiment, therefore, MAS also increases the resolution in the ω_2 dimension for nutation spectroscopy. However, since MAS produces echoes at times $t = n/\nu_{rot}$, where ν_{rot} is the frequency of rotation and n any integer, it can disturb the nutation experiment for long pulses. It was shown [92] that in order to avoid destructive interference of the spinning-dependent phase of the magnetization the relation $t_1 \leqq 1/4\omega_{rot}$ must be fulfilled, if the comparison is made with static nutation spectra. Nielsen et al. [36] presented recently the theoretical treatment and computer simulations of nutation MAS NMR spectra and obtained good agreement with experimentally obtained spectra. They demonstrated that MAS nutation spectra are similar to static nutation spectra with a lower ω_Q/ω_{rf} ratio, e.g. the spectrum with $\omega_Q/\omega_{rf} = 2$, $\omega_{rot} = 10\,\text{kHz}$ is similar to the spectrum $\omega_Q/\omega_{rf} = 1$, $\omega_{rot} = 0\,\text{kHz}$ [36].

The spin–lattice relaxation, T_1, is an important parameter for the nutation experiment. If the recycle delay does not allow all spins to relax, the build up of the z-magnetization is incomplete before the next pulse is applied. Since one starts with a pulse for which $\omega_{rf}t_1 \ll \pi/2$ the intensity after the next pulse with increased t_1 will be reduced due to incomplete spin–lattice relaxation which has a similar effect than a larger ω_Q. It was shown [85] that in the course of the whole 2D-experiment there appear overtones of ω_{rf} in the ω_1 direction which can be easily mistaken for components with larger ω_Q.

$T_{2\rho}$ spin–spin relaxation in the rotating frame may affect the width of the line in ω_1 direction, if $T_{2\rho} \lesssim t_1$, as does $T_{1\rho}$ or $T_{2\,\mathrm{effective}}$ in the solid-state spectra narrowed by multiple-pulse groups. The maximum linewidth of the nutation line or a minimum in $T_{2\rho}$ is found if fluctuation rates in the lattice are about equal to the splitting in the rotating frame. Janssen et al. [86] found such a behavior of the width of the nutation line for the hydrated zeolite NaA as a function of the temperature.

4.3 *Advanced Concepts in Nutation Spectroscopy*

The standard nutation experiment, Fig. 4.4A, consisting of only one pulse of variable length t_1, was combined with rotary echoes by Veeman and coworker [86, 91] (cf. Fig. 4.4B and C), for rotary echoes cf. Abragam [1] pp. 70–71, and with a second selective $\pi/2$ pulse (cf. Fig. 4.4D) by Samoson and Lippmaa [92]. Nielsen et al. [36] detect the multiple-quantum coherences, which were produced by the first pulse if ω_Q and ω_{rf} are in the same order of magnitude, with a mixing pulse of fixed duration and proper phase cycling to ensure transfer of selected pQ coherences (cf. Fig. 4.4E).

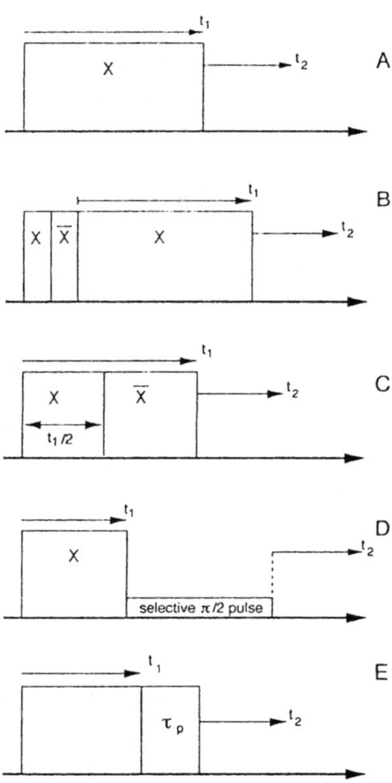

Fig. 4.4. Pulse sequences for nutation experiments. **A:** one pulse of variable length t_1, **B:** nutation experiment with two preceded pulses for rotary echoes [86], **C:** the combined rotary echo and nutation experiment [91] **D:** nutation experiment combined with a second selective $\pi/2$ pulse [92], **E:** multiple-quantum MAS nutation [36] with the first pulse of variable length t_1 and a second mixing pulse of fixed duration τ_p and proper phase cycling to ensure transfer of selected pQ coherences

In the rotary echo nutation experiment, nuclei with different rotating frame spin–spin relaxation time $T_{2\rho}$ can be distinguished, and information about relaxation effects can be obtained. In the first version of this technique, shown in Fig. 4.4B, the nutation pulse is preceded by two pulses of the same length τ, but of opposite phase [86]. This procedure refocuses the magnetization at the time 2τ along the z-axis for $\omega_Q \ll \omega_{rf}$ and $\omega_Q \gg \omega_{rf}$, but not for the intermediate case. For nutation spectra with large intensities at frequencies ω_{rf} and $(I + 1/2)\omega_{rf}$ the intensity of the signal is reduced by a factor of about $\exp\{-2\tau/T_{2\rho}\}$. $T_{2\rho}$ depends on the correlation time τ_c describing the orientation of the quadrupole tensor with respect to the external magnetic field. The minimum value of $T_{2\rho}$, which is slightly bigger than the reciprocal static linewidth, is reached if the correlation time is about equal to the reciprocal static linewidth.

For the $X\bar{X}$-nutation experiment introduced by van der Mijden et al. [91] the nutation pulse itself is replaced by a rotary echo (cf. Fig. 4.4C), and therefore, no signal will be observed from a spin system which has either an extremely small or extremely large quadrupole interaction. The spectrum of a spin system with an intermediate quadrupole interaction $\omega_Q \approx \omega_{rf}$ gives nutation frequencies different from those calculated for the standard experiment and the spectrum of the central transition contains negative intensities [91]. Whereas the influence of the preceded rotary echo on the signals at $\omega_1 = \omega_{rf}$ and $\omega_1 = (I + 1/2)\omega_{rf}$ can be discussed qualitatively, the general interpretation of $X\bar{X}$-nutation spectra requires the comparison with a set of simulated $X\bar{X}$-nutation spectra to obtain quantitative results.

Negative nutation frequencies of the central transition are manifested by negative values of the z-component polarization. Those non-observable magnetizations can be converted into the observable transverse magnetization (coherence) with an additional $\pi/2$ pulse attenuated in amplitude, so that it is selective for the central transition. Samoson and Lippmaa [92] presented nutation spectra with the y- and z-components of central transition magnetization of $^{23}NaNO_3$. For a ratio $\omega_Q/\omega_{rf} = 0.15$ they showed an ordinary $I = 3/2$ spectrum, positive central line and positive satellites, and a spectrum with negative central line and positive satellite lines for the y- and z-components, respectively.

Nielsen et al. [36] applied a two-quantum and zero-quantum phase cycling of the mixing pulse in order to transfer the 2MQ coherences into observable 1MQ coherences and to analyze the functional behavior of the longitudinal multiple order magnetization (LMO), respectively. They noted that the transfer into MQ coherences highly depends on ω_Q/ω_{rf} and ω_{rot} [36].

5 Multiple-Pulse Techniques

Among the multiple-pulse techniques for the investigation of quadrupole nuclei in powders the (two-pulse) spin-echo methods are the most important. Since a wide distribution of resonances is to be expected for quadrupole nuclei, spin-echo methods are often inevitable for the recording of quadrupolar broadened NMR signals. Before we will review the spin-echo methods for different strengths of quadrupole couplings, the two-pulse free induction decay method [94, 95] is discussed since it reveals excitation phenomena for quadrupole nuclei when observing only the central transition, and thus, it is of importance for spin-echo measurements on powders. The two-pulse free induction decay method which, from the point of view of applications, is rather comparable with the nutation spectroscopy, uses nonselective (partly selective) pulses for the measurement of the quadrupole parameters and shows in general the influence of partly selective excitation on any multiple pulse intensity measurement. Therefore, the two-pulse free induction decay method will be explained in more detail before the spin-echo techniques. Cross-polarization is the subject of Sect. 6.2. The Spin Echo Double Resonance (SEDOR) technique, introduced by Emshwiller et al. [96], has found only a few applications for quadrupole nuclei, hence we refer to the original literature of Lang et al. [97, 98].

For further discussion we introduce the basic definitions for any two-pulse experiment. With the definitions given in Fig. 5.1, the observed NMR signal, $E(t_1, t_2)$, at the time t_2 reads in the superoperator notation, cf. Sect. 1.7.

$$E(t_1,t_2)=\left\langle I_- \left| \exp\left\{-\frac{\mathrm{i}}{\hbar}\mathscr{H}_2 t_2\right\} \exp\left\{-\frac{\mathrm{i}}{\hbar}\mathscr{H}_3 \tau_2\right\} \exp\left\{-\frac{\mathrm{i}}{\hbar}\mathscr{H}_2 t_2\right\}\right.\right.$$
$$\left.\left. \cdot\exp\left\{-\frac{\mathrm{i}}{\hbar}\mathscr{H}_1 \tau_1\right\}\right| I_z \right\rangle, \qquad (5.01)$$

where, the factor for normalization, $1/\langle I_x | I_x \rangle$ has been dropped.

For hard pulses, $\|\mathscr{H}_{rf}\| \gg \|\mathscr{H}_Q\|$, the influence of the quadrupole interaction and a small resonance offset $\Delta\omega$ can be omitted during pulsing. Then, the Hamiltonians in Eq. (5.01) are

$$\mathscr{H}_1 = \hbar\omega_{rf} \exp\{-\mathrm{i}\phi_1 I_z\} I_y \exp\{+\mathrm{i}\phi_1 I_z\} \qquad (5.02)$$

$$\mathscr{H}_2 = \frac{\hbar\omega_Q'}{6}(3I_z^2 - I(I+1)) + \Delta\omega I_z \qquad (5.03)$$

$$\mathscr{H}_3 = \hbar\omega_{rf} \exp\{-\mathrm{i}\phi_2 I_z\} I_y \exp\{+\mathrm{i}\phi_2 I_z\} \qquad (5.04)$$

cf. Eqs. (1.55) and (1.57). Inserting the Hamiltonians into Eq. (5.01), and, by

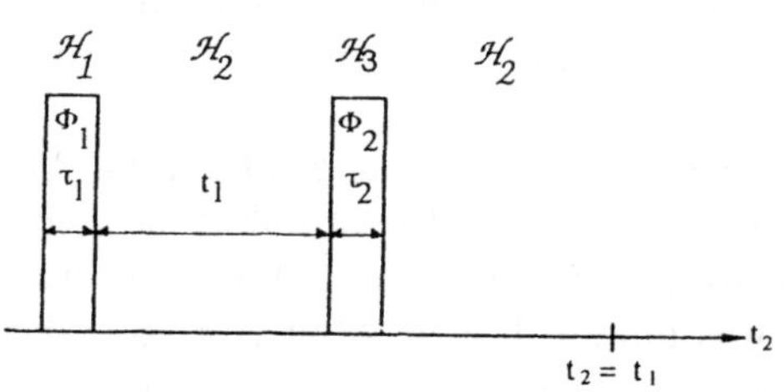

Fig. 5.1. Sketch of the two-pulse experiment. $\mathscr{H}_{1,2,3}$ represents the Hamiltonian in the interaction representation, $\Phi_{1,2}$ and $\tau_{1,2}$ are the phases and the duration of the two pulses, respectively

defining $k_x = \cos \Phi_1 \cdot \sin(\omega_{rf}\tau_1), k_y = \sin(\omega_{rf}\tau_1) \cdot \sin \Phi_1, k_z = \cos(\omega_{rf}\tau_1)$ we have

$$E(t_1, t_2) = \sum_{j=x,y,z} k_j \sum_{m,m',m''=-I}^{I} 2W_m \exp\{iA\} \exp\{i(m' - m'' + 1)\Phi_2\}$$
$$\cdot d^{(I)}_{mm'}(\omega_{rf}\tau_2) d^{(I)}_{m''m+1}(-\omega_{rf}\tau_2)\langle m'|I_j|m''\rangle, \quad (5.05)$$

where, the dependence on t_1 and t_2 is contained in

$$A = \tfrac{1}{2}\omega'_Q[(2m + 1)t_2 + (m''^2 - m'^2)t_1] + \Delta\omega[t_2 + (m'' - m')t_1], \quad (5.06)$$

and

$$d^{(I)}_{m'm}(\beta) = \langle m'|\exp\{-i\beta I_y\}|m\rangle, \quad (5.07)$$

are the Wigner rotation matrices, cf. Table 1.1. For the first sum in Eq. (5.05), $j = z$ describes the action of the second pulse on I_z which is simply the FID for a nonselective excitation and will not be considered any further.

5.1 *Two-Pulse Free Induction Decay*

The information of the first-order quadrupole interaction can be transformed onto the central transition intensity by two pulses as first reported in Ref. [94]. The procedure is most useful for powders where the first-order frequency spread of the satellites cannot be detected by other means and second-order effects are obscured by other interactions.

Supposed, the quadrupole nucleus, initially in thermal equilibrium, is excited by a hard y-pulse with $\omega_{rf}\tau \ll 1$. Then, immediately after the pulse there appear $2I$ components of magnetization, polarized in the x-direction. In the rotating frame these $2I$ components have a resonance offset $\Delta\omega$ due to the first-order quadrupole interaction: $\Delta\omega_m = (2m + 1)\omega'_Q$, where m runs from $-I$, $-I + 1, \ldots,$ $I - 1$, cf. Sect. 2.2, Eq. (2.18). The x-components of these polarizations vary with $\cos[(2m + 1)\omega'_Q t]$. A second hard y-pulse at time $t = t_1$ creates, a new x-component formed by spins which were still in z-direction after the first pulse. Then, with respect to the polarizations created by the first pulse, it affects only spins polarized in x-direction, the amount of which is proportional to $\cos[(2m + 1)\omega'_Q t_1]$, at a time $t = t_1$. Hence, depending on the time delay between the pulses, a varying number of spins forming the satellite transitions will be transferred into the central transition. Since these contributions are sinusoidal in $\omega'_Q t_1$ the intensity of the central transition is modulated by $\cos[(2m + 1)\omega'_Q t_1]$. A second Fourier transform, with respect to t_1, of the dependence of the intensity of the central transition on the pulse length t_1 should give definite contributions at frequencies $\omega = (2m + 1)\omega'_Q$.

A theoretical analysis of this effect has been reported by Haase in Refs. [94, 95] for the case of nonselective pulses, the phase shifting between pulses has been discussed, and information about the quadrupole parameter of ^{27}Al in powdered alum, corundum, and low-cristobalite was obtained. P. P. Man described the experiment for $I = 3/2$ with the fictitious spin-1/2 operator formalism using two in-phase rf pulses with a short delay between them and variable length of the second pulse [88, 99], and in addition, with alternating pulse phase and alternating receiver phase [100]. The quadrupole coupling of ^{23}Na in powdered $NaNO_3$ [88], of ^{7}Li in a single crystal of $LiTaO_3$ [99, 100] were studied by the dependence of the central line intensity on the width of the second rf pulse.

The intensity of the central transition $F_c(t_1)$ after two pulses of arbitrary phase follows from Eqs. (5.05–5.07), by choosing $t_2 = 0$, $m = -1/2$, and, using the symmetry properties of the Wigner matrices. The resonance offset is being neglected.

$$\begin{aligned} F_c(t_1) = {} & 2W_{-1/2}^2[k_z \sin(\omega_{rf}\tau_2)\cos(\omega_{rf}\tau_1)\exp\{i\Phi_2\} \\ & + k_- F_{-1/2}\exp\{i2\Phi_2\} + k_+ G_{-1/2}] \\ & + 2W_{-1/2} \sum_{m'>0} 2W_{m'}[k_- \exp\{i2\Phi_2\}F_{m'} + k_+ G_{m'}] \\ & \cdot\cos\{\omega_Q'(m'+1/2)t_1\}, \end{aligned} \tag{5.08}$$

where

$$k_+ = k_x + ik_y, \quad k_- = k_x - ik_y, \quad k_x = \cos\Phi_1 \sin(\omega_{rf}\tau_1), \quad k_y = \sin(\omega_{rf}\tau_1)\sin\Phi_1,$$
$$F_{m'} = d_{1/2,m'}(\omega_{rf}\tau_2)d_{-1/2,m'+1}(\omega_{rf}\tau_2), \quad G_{m'} = d_{-1/2,m'}(\omega_{rf}\tau_2)d_{1/2,m'+1}(\omega_{rf}\tau_2).$$

For the special case of equal rf pulses, $\tau_1 = \tau_2$, $\Phi_1 = \Phi_2 = 0$, Eq. (5.08) can be written as

$$F_c = F_0 + \sum_{m>0}^{I-1} F_{m+1/2}\cos\{(m+1/2)\omega_Q' t_1\}, \tag{5.09}$$

where

$$\begin{aligned} F_0 &= W_{-1/2}^2 \sin(2\omega_{rf}\tau) + 2W_{-1/2}^2 C_{-1/2}\sin(\omega_{rf}\tau), \\ F_{m+1/2} &= 4W_{-1/2}W_m C_m \sin(\omega_{rf}\tau), \quad C_m = F_m + G_m. \end{aligned} \tag{5.10}$$

It can be seen from Eq. (5.09) that for powders a Fourier transform with respect to t_1 yields the first-order powder pattern. The first term of Eq. (5.09) represents a contribution from the central transition only, not influenced by the satellites, which gives rise to a quasi central transition after the Fourier transform with respect to t_1, modified in the intensity. The coefficients C_m in Eq. (5.10) read

for $I = 3/2$

$$C_{-1/2} = \frac{1}{16}[7\cos(\omega_{rf}\tau) + 9\cos(3\omega_{rf}\tau)], \tag{5.11}$$

$$C_{1/2} = \frac{3\sqrt{3}}{16}[\cos(\omega_{rf}\tau) - \cos(3\omega_{rf}\tau)],$$

for $I = 5/2$

$$C_{-1/2} = \frac{1}{64}[18\cos(\omega_{rf}\tau) + 21\cos(3\omega_{rf}\tau) + 25\cos(5\omega_{rf}\tau)],$$

$$C_{1/2} = \frac{\sqrt{2}}{128}[22\cos(\omega_{rf}\tau) + 3\cos(3\omega_{rf}\tau) - 25\cos(5\omega_{rf}\tau)],$$

$$C_{3/2} = \frac{\sqrt{5}}{128}[10\cos(\omega_{rf}\tau) - 15\cos(3\omega_{rf}\tau) + 5\cos(5\omega_{rf}\tau)]. \tag{5.12}$$

The dependence of the components F_m on the flip angle $\omega_{rf}\tau$, Eq. (5.09) and (5.10), for $I = 7/2$ are shown in Fig. 5.2. As can be seen, for small flip angles, i.e. for short pulses, there will be negligible contributions from the outer satellites, and, one is left with a reduced spin-3/2 powder pattern which will show only the next neighbored satellites.

Since one is interested in the variation of the central line intensity due to satellite contributions, folded over by the second pulse, the constant term should be minimized. It has been shown [95] that the most sufficient phase shifting is $\Phi_1 = 0$, $\Phi_2 = 0$; $\Phi_1 = 0$, $\Phi_2 = \pi$, which reduces the constant term by a factor of two, but, it does not increase the contributions from the outer satellites ($I > 3/2$).

Compared with the nutation spectroscopy the two-pulse free induction decay method seems to be much easier since a second Fourier transform yields the well known powder pattern, and, the quadrupole parameters can be deduced directly. Another advantage of the two-pulse method is its insensitivity to ω_{rf} inhomogeneity, as can be seen in Fig. 5.2. Nevertheless, the two-pulse method meets the same difficulties with respect to the excitation, as does the nutation spectroscopy: the bandwidth of excitation has to be bigger than the first order quadrupole coupling. Although the two-pulse free induction decay method yields predictable lineshapes (in the second dimension) for even a partly selective excitation, the delay between the two pulses becomes the critical parameter rather than the smallest pulse width for the nutation spectroscopy. For both methods the central transition must be observed selectively. The appearance of an echo for the two-pulse technique can influence the results if larger flip angles are used.

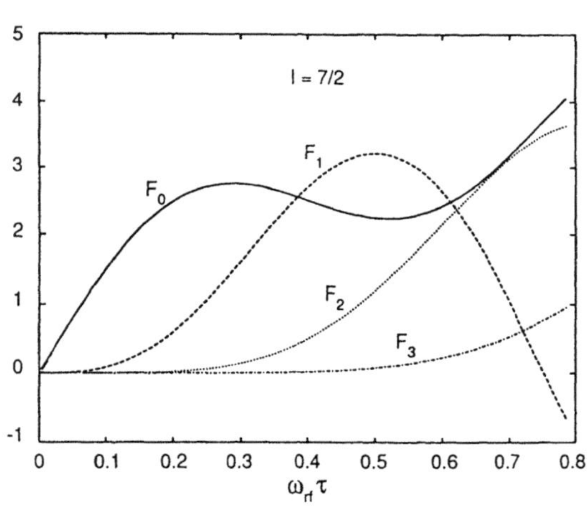

Fig. 5.2. Intensity of the components F_m on the flip angle of one of two identical pulses, $I = 7/2$

5.2 Echoes

The refocusing effect of a π pulse, applied some time after a preceding $\pi/2$ pulse, was discovered by E. L. Hahn [101]. Echo techniques are mainly used for two reasons: First, by varying the pulse distance the resulting envelope of the echo decay gives additional information about the spin system. Second, a loss of signal during the ring down of the probe and the recovery of the receiver can be minimized. A quadrupole interaction strongly influences the formation of an echo: Homonuclear dipolar interactions which mainly cause the decay of the spin-echo amplitude become less effective since spin-flipping between different transitions is forbidden. But, due to the nature of the quadrupole coupling, which is in first-order proportional to I_z^2, a refocussing is not complete. Furthermore, a limited range of excitation causes a complicated spin-echo behavior, especially for powder samples.

The majority of the previous echo studies on quadrupole nuclei used non-selective excitation and did not involve powdered substances. Das and Saha [102] calculated the echo response of nuclei experiencing first-order quadrupole interactions to a sequence of two pulses of identical duration and phase, neglecting dipole interactions. Solomon [103] showed that, when quadrupole interactions cause significant evolution of the spin system during the pulses, forbidden echoes occur which are not bell-shaped curves like the allowed echoes, but, rather the derivatives of such curves [1], named "sine" echoes. Butterworth [104] observed the superposition of an echo at $t = 2\tau$ arising from magnetic interaction (inhomogeneity of the external field) with the quadrupole echo. The advantage of different phases of the $\pi/2$ pulses for the first-order quadrupole echoes was demonstrated for ^{79}Br and ^{81}Br in alkali crystals [105], ^{131}Xe in solid Xe [106] and for spin-5/2 nuclei [107]. Abe et al. [108] calculated analytically the oscillations of the echo amplitude for $I = 3/2$ and the case that the quadrupole interaction is either smaller or larger than the interaction with the rf pulses. Numerical computations covering also the intermediate case were performed by Sobral et al. [109]. Mehring and Kanert [110] discussed the echo amplitude as a function of the angle of the second of two in-phase pulses and analyzed the lineshape of the spin echo for $I = 3/2$ through $I = 9/2$. Both, quadrupole and magnetic broadening were taken into account for the echoes observed in vanadium compounds by Schoep et al. [111]. The first three-pulse sequence was analyzed to study cross-relaxation effects of $I = 3/2$ nuclei by Mansfield et al. [112]. A comprehensive treatment of all three-pulse sequences was performed by Halstead and Sanctuary et al. [113] using the multipole approach [6, 114] for spin-3/2 nuclei subjected to "hard" rf pulses and an inhomogeneous distribution of quadrupole interactions and a local dipolar field. The concept of multipoles was also used in order to explain distinct relaxation times for the higher rank polarizations in single crystals of KI (^{127}I, spin 5/2) [115]. Density matrix solutions for sequences of two "soft" pulses on $I = 3/2$ with $\eta = 0$ were given by Campolieti et al. [116]. Studies of a spin-3/2 system by an echo sequence without limitation to "hard" or "soft" pulses were performed by Man, who extended the two-pulse

FID technique with varying length of the second pulse, cf. Ref. [100], to echo studies [117, 118].

Echoes in magnetically ordered substances were investigated by several authors [119–121]. Furo and Halle [32, 122, 123] developed the two-dimensional quadrupole echo method for nuclei in anisotropic liquids with small quadrupole splittings. They also included powder samples [122].

Mansfield [74] considered the selective excitation of the central transition and treated the echo after a sequence of two identical pulses under the influence of a homonuclear dipole interaction. Later a spin-echo Fourier transform NMR technique was used to obtain undistorted shapes of the quadrupole broadened central lines [124]. The "two-dimensional homonuclear separation on interaction method" [27] could be applied to separate the homonuclear dipolar interaction (ω_1) and the inhomogeneous quadrupole interaction (ω_2) [125]. Both studies [124, 125] and recent treatments of Haase and Oldfield [75] and Furó and Halle [126, 127] take into consideration that only the central transition is excited and that the second-order quadrupole broadening of the observed central line is in order of magnitude of the dipole broadening. This approach became very important for the study of quadrupole nuclei in inorganic solids and will be discussed in Sect. 5.2.5 in detail.

5.2.1 Hard and Nonselective Pulses Without Dipole Interaction

For hard and nonselective pulses it follows from Eq. (5.05) that if $\Delta\omega = 0$, the influence of ω'_Q vanishes for

$$(2m + 1)t_2 + (m''^2 - m'^2)t_1 = 0. \tag{5.13}$$

Several echoes at times $t_2 = kt_1$ are possible [103]. For $I = 3/2$ and $I = 5/2$ we find from Eq. (5.13) echoes for $k = 1/2, 1, 3/2$ and $k = 1/2, 1, 3/2, 2, 3$, respectively. Inspection of the matrix elements $\langle m'|I_j|m''\rangle$, Eq. (5.05), reveals that for $j = x, y$ the values for m', m'' are restricted to $|m'' - m'| = 1$ yielding the allowed echoes: $k = 1$ for $I = 3/2$; $k = 1/2, 1, 2$ for $I = 5/2$.

From Eq. (5.06) it follows that, as long as $\Delta\omega$ is the same for all nuclei of the powder each Solomon-echo shows a phase factor. However, an anisotropy of the chemical shift or an inhomogeneous dipole interaction makes $\Delta\omega \neq 0$, and, the contributions of all nuclei will cancel each other. From Eq. (5.06) it can be seen,

$$t_2 = \frac{\frac{1}{2}\omega'_Q(m' + m'') + \Delta\omega}{\frac{1}{2}\omega'_Q(2m + 1) + \Delta\omega}(m' - m'')t_1, \tag{5.14}$$

that if $\Delta\omega$ is not the same for all nuclei, an echo forms when $(2m + 1) = (m' + m'')$. Thus, only the echo at $t_2 = t_1$ remains. This is the reason why for the study of quadrupole echoes at times $t_2 \neq t_1$ oriented material must be investigated.

The amplitude of the echoes can be calculated from Eq. (5.05) with the help of the Wigner rotation matrices. As was shown by Bonera [105] and Warren [106], the maximum echo is obtained for $\Phi = 90°$ and $\omega_{rf}\tau_1 = \omega_{rf}\tau_2 = \pi/2$.

Introducing $t_2 = t + k \cdot t_1$ in Eq. (5.05), the dependence of $E(t, t_1, t_2)$ on t yields the shape of the k-th echo. The leading term, $\exp\{[\frac{1}{2}\omega'_Q(2m+1) + \Delta\omega]t\}$, shows the echo due to the different transitions, and that it is equivalent to the FID.

5.2.2 *Soft Pulses Without Dipole Interaction*

For soft pulses, $\|\mathscr{H}_{rf}\| \approx \|\mathscr{H}_Q\| \gg \|\mathscr{H}_D\|$, the influence of the quadrupole interaction during the pulse has to be considered, cf. Sect. 2.3.

As a consequence of soft, partly selective excitation, the density operator after the first pulse is no longer proportional to I_x or I_y, and, the selection rule, $|m' - m''| = 1$, following from $\langle m'|I_j|m''\rangle$, Eq. (5.05), is no longer valid. In other words, the spectral energy density of excitation for satellite transition with a large frequency offset may be very weak, so that in Eq. (5.05) $\langle m|I_y|m-1\rangle \approx 0$. Thus, additional echoes, derived from Eq. (5.13), may now be observed. These are called *Forbidden Echoes*, cf. Ref. [1, 103].

5.2.3 *Dipole Interaction for First-Order Quadrupole Echoes*

The heteronuclear dipole interaction, which is proportional to I_zS_z, can generally by removed by the application of a second π-pulse. However, a strong coupling amongst the S-spins, as well as a short spin–lattice relaxation of the S-spins, can destroy the formation of the I-spin echo.

If the homonuclear dipole interaction is large compared to the quadrupole interaction, $\|\mathscr{H}_D\| \gg \|\mathscr{H}_Q\|$, its influence is essentially the same as for spin-1/2 nuclei [74]. If the homonuclear dipole interaction is as strong as the quadrupole interaction, $\|\mathscr{H}_D\| \approx \|\mathscr{H}_Q\|$, in dependence on the orientation of the EFG tensor with respect to the static magnetic field, the spin flipping may be restricted to equivalent nuclei and a single transition. A detailed investigation for this range of homonuclear/heteronuclear dipole and quadrupole coupling has not yet been done.

If the quadrupole interaction is large compared to the dipole interaction, $\|\mathscr{H}_Q\| \gg \|\mathscr{H}_D\|$, spin exchange due to homonuclear dipole interaction between adjacent levels can be considered as being suppressed, completely, and, the echo decay can be investigated by selective excitation of each transition, separately.

5.2.4 *Selective Excitation of a Single Transition Without Dipole Interaction*

If $\|\mathscr{H}_Q\| \gg \|\mathscr{H}_{rf}\| \gg \|\mathscr{H}_D\|$ one can safely tune to any transition for single crystals or mainly excite the central transition for powders. This can be described by the reduced spin-1/2 formalism [1].

As outlined in Sect. 2.4, cf. Eqs. (2.22–2.24), by tuning to the appropriate transition, and, omitting operators proportional to the unity operator one has merely to replace,

$$I_z \rightarrow aS_z, \quad I_z^2 \rightarrow (2m+1)S_z, \quad I_y \rightarrow 2W_m S_y. \tag{5.15}$$

Since one is interested in the intensity of a single transition the analogue to Eq. (5.05) is

$$\begin{aligned} E(t_1,t_2) = \langle S_- | S_z \rangle = \langle S_- | &\exp\{-\mathrm{i}\Omega S_z t_2\} \exp\{-\mathrm{i}\Phi_2 S_z\} \\ &\cdot\exp\{-\mathrm{i}2W_m\omega_{rf}\tau_2 S_y\} \exp\{\mathrm{i}\Phi_2 S_z\} \exp\{-\mathrm{i}\Omega S_z t_1\} \\ &\cdot\exp\{-\mathrm{i}\Phi_1 S_z\} \exp\{-\mathrm{i}2W_m\omega_{rf}\tau_1 S_y\} \exp\{\mathrm{i}\Phi_1 S_z\} | S_z \rangle, \end{aligned} \tag{5.16}$$

where, Ω represents some remaining resonance offset, e.g. due to second-order quadrupole interaction. From Eq. (5.16) it follows for a pair of in-phase 90°–180° selective pulses, i.e. for which now $2W_m\omega_{rf}\tau_1 = \pi/2$, $2W_m\omega_{rf}\tau_2 = \pi$, and $\Phi_1 = \Phi_2 = 0$,

$$\begin{aligned} E(t_1,t_2) &= \langle S_- | \exp\{-\mathrm{i}\Omega S_z t_2\} \exp\{-\mathrm{i}\pi S_y\} \exp\{-\mathrm{i}\Omega S_z t_1\} \\ &\qquad \cdot\exp\{+\mathrm{i}\pi S_y\} \exp\{-\mathrm{i}\tfrac{3}{2}\pi S_y\} | S_z \rangle \\ &= \langle S_- | \exp\{-\mathrm{i}\Omega S_z t_2\} \exp\{+\mathrm{i}\Omega S_z t_1\} | -S_x \rangle, \end{aligned} \tag{5.17}$$

an echo is observed at $t_2 = t_1$ in full analogy to spin-1/2 nuclei subjected to an inhomogeneous interaction.

5.2.5 Selective Excitation of a Single Transition With Dipole Interaction

Since the calculation of the spin-echo amplitude in the presence of dipolar interactions can be rather complicated, the spin-echo decay will be discussed in terms of the second moment spin-echo decay, following the work by Haase and Oldfield [75]. This consideration of homonuclear and heteronuclear dipolar interactions is necessary in order to explain the spin-echo behavior of a large amount of inorganic compounds [75].

Using the interaction representation, transformed by the Zeeman interaction, one can describe the intensity of an echo occurring at times 2Δ, as indicated in Fig. 5.1 with $t_1 = t_2 = \Delta$, by the following trace:

$$E(2\Delta) = \mathrm{tr}\{\rho(2\Delta) I_+\} \equiv \langle I_- | \rho(2\Delta) \rangle, \tag{5.18}$$

where

$$\begin{aligned} |\rho(2\Delta)\rangle = &\exp\{-\mathrm{i}\mathcal{H}_2\Delta\} \exp\{-\mathrm{i}\mathcal{H}_3\tau_2\} \exp\{-\mathrm{i}\mathcal{H}_2\Delta\} \\ &\cdot\exp\{-\mathrm{i}\mathcal{H}_1\tau_1\} |\rho_0\rangle. \end{aligned} \tag{5.19}$$

For convenience, we use the following transformation:

$$\mathcal{H}_{2r} = \exp\{-\mathrm{i}\mathcal{H}_3\tau_2\} \mathcal{H}_2 \exp\{\mathrm{i}\mathcal{H}_3\tau_2\} \tag{5.20}$$

and can then write:

$$\begin{aligned} |\rho(2\Delta)\rangle = &\exp\{-\mathrm{i}\mathcal{H}_2\Delta\} \exp\{-\mathrm{i}\mathcal{H}_{2r}\Delta\} \exp\{-\mathrm{i}\mathcal{H}_3\tau_2\} \\ &\cdot\exp\{-\mathrm{i}\mathcal{H}_1\tau_1\} |\rho_0\rangle. \end{aligned} \tag{5.21}$$

where the Hamiltonians $\mathscr{H}_1(\mathscr{H}_3)$ and $\mathscr{H}_2$, all expressed in the interaction representation, describe the interactions during the first (second) pulse and in the absence of rf excitation, respectively, and may be approximated by means of the Magnus expansion, as

$$\mathscr{H}_2 = \bar{\mathscr{H}}_Q^{(1)} + \bar{\mathscr{H}}_Q^{(2)} + \bar{\mathscr{H}}_D^{(1)} \quad \text{and} \quad \mathscr{H}_{1(3)} = \bar{\mathscr{H}}_{rf1(3)}^{(1)} + \bar{\mathscr{H}}_Q^{(1)}. \tag{5.22}$$

Since the quadrupolar interaction dominates, and the interaction of the spins with the rf field, $\mathscr{H}_{rf}$, is assumed to exceed second-order quadrupolar contributions as well as the dipolar interactions amongst the spins, one retains during the rf excitation only the first-order quadrupolar term, $\bar{\mathscr{H}}_Q^{(1)}$, and the first-order part of the rf excitation, $\bar{\mathscr{H}}_{rf}^{(1)}$. In the absence of rf excitation, one considers the quadrupolar effects in first-order, as well as in second-order, $\bar{\mathscr{H}}_Q^{(1)} + \bar{\mathscr{H}}_Q^{(2)}$, and the dipolar interaction in first-order, $\bar{\mathscr{H}}_D^{(1)}$. Finally, in the high temperature approximation, it is

$$|\rho_0\rangle \sim \sum_j |I_z^j\rangle. \tag{5.23}$$

In order to determine the spin-echo decay behavior, the second moment, M_{2E}, of the envelope of a train of echoes is calculated, as:

$$M_{2E} = -\frac{d^2}{d(2\Delta)^2} E(2\Delta)\bigg|_{2\Delta=0}. \tag{5.24}$$

Introducing Eqs. (5.18–5.23) into Eq. (5.24), one finds:

$$M_{2E} = \tfrac{1}{4}\{\langle I_- | [\mathscr{H}_2, [\mathscr{H}_2, \rho_1]]\rangle + 2\langle I_- | [\mathscr{H}_2, [\mathscr{H}_{2r}, \rho_1]]\rangle$$
$$+ \langle I_- | [\mathscr{H}_{2r}, [\mathscr{H}_{2r}, \rho_1]]\rangle\}, \tag{5.25}$$

where

$$|\rho_1\rangle = \exp\{-i\mathscr{H}_3\tau_2\} \exp\{-i\mathscr{H}_1\tau_1\} |\rho_0\rangle. \tag{5.26}$$

This calculation of the second moment of the spin-echo decay, M_{2E}, is relatively easy for the case of a dominating homonuclear dipolar interaction, assuming all nuclei are subjected to the same quadrupole coupling. In Ref. [75] it is shown that M_{2E} for each transition, selectively excited by a $\pi/2 - \pi$ pulse sequence can be expressed by the following Equations:

$$M_{2E} = E_L^{(n)} \gamma^4 \hbar^2 \sum_j b_{ij}^2, \tag{5.27}$$

where $E_L^{(n)}$ is a spin-dependent factor, "L" refers to "like spins" (identical parameters for the quadrupole coupling) and the label $n = 0, 1, \ldots$ $(n < (I + 1/2))$ denotes the central transition and the first satellite transitions, respectively. For the central transition, one finds

$$E_L^{(0)} = \frac{2}{9(2I+1)}(1 + 4W_{-1/2}^2 + 4W_{-1/2}^4 + W_{+1/2}^4), \tag{5.28}$$

where W_m is defined in Eq. (2.19). In Eq. (5.28), "1" represents the probability that neither spin changes its quantum number, due to $\bar{\mathscr{H}}_D^{(1)}$, until the echo forms. $W^2_{-1/2}$ corresponds to transitions where, until application of the second pulse, spin flipping occurs between the central levels, and, between the second pulse and the echo there is no flipping; $W^4_{-1/2}$ corresponds to flip–flops during both time periods; and finally, $W^4_{+1/2}$ stands for flip–flops in the transitions next to the central one ($\pm 3/2$) during both time periods.

Equation (5.27) has to be compared with the expression for the second moment obtained from the FID, cf. Eq. (3.19), which can be written as

$$F_L = \frac{2}{9(2I+1)}\left[2\sum_{m=-I}^{I} m^2 + 4W^2_{-1/2} + 4W^4_{-1/2} + 4W^4_{1/2}\right]. \tag{5.29}$$

Comparison with Eq. (5.28) shows that, except for terms with $m = \pm 1/2$, the diagonal contribution cancels for the echo, i.e. spins with $|m| > 1/2$ which do not flip until the echo forms do not contribute to the echo decay. Also, the flip–flop term between the levels next to the central ones is reduced by a factor of 4, but, its contribution in both Eq. (5.28) and (5.29) is negligible. Inserting spin quantum numbers, it can be seen from Eqs. (5.28) and (5.29) that the ratios $E_L/F_L \approx 0.5 \pm 0.1$, and thus essentially independent of the spin-quantum number, I.

As might have been anticipated, the Hahn spin-echo decay behavior of the "isolated" central transition is mainly governed by the dipolar interactions between the fraction of spins having magnetic quantum numbers of $m = \pm 1/2$.

For the selective excitation of satellite transitions similar relations exist. Reference [75] gives expressions for $I = 5/2$ from which the generalization to arbitrary spin quantum number becomes apparent.

For the first satellites:

$$F_L^{(1)} = \frac{2}{9(2I+1)}\left[2\sum_{m=-I}^{I} m^2 + 4W^2_{+1/2} + 4W^4_{+1/2} + 2W^4_{-1/2} + 2W^4_{+3/2}\right], \tag{5.30}$$

$$E_L^{(1)} = \frac{2}{9(2I+1)}\left[5 + 4W^2_{+1/2} + 4W^4_{+1/2} + \tfrac{1}{2}W^4_{-1/2} + \tfrac{1}{2}W^4_{+3/2}\right]. \tag{5.31}$$

For the second satellites:

$$F_L^{(2)} = \frac{2}{9(2I+1)}\left[2\sum_{m=-I}^{I} m^2 + 4W^2_{+3/2} + 4W^4_{+3/2} + 2W^4_{+1/2}\right], \tag{5.32}$$

$$E_L^{(2)} = \frac{2}{9(2I+1)}\left[14 + 4W^2_{+3/2} + 4W^4_{+3/2} + \tfrac{1}{2}W^4_{+1/2}\right]. \tag{5.33}$$

For $I = 5/2$, intensity ratios are:

$$\frac{F_L}{E_L} = \frac{321}{137};\quad \frac{F_L^{(1)}}{E_L^{(1)}} = \frac{289}{128};\quad \frac{F_L^{(2)}}{E_L^{(2)}} = \frac{217}{121}. \tag{5.34}$$

If like spins experience a slightly different quadrupolar coupling, i.e. the first-order quadrupolar interactions are different (such spins have also been called "semi-like" spins, cf. Abragam [1], p. 130), their central transition frequencies are still the same, but the satellite transition frequencies can be very different. In this case it can be concluded from the above analysis that the spin-echo decay for the central transition will not be changed (only the very small contribution from spin-flipping in the neighbored transitions vanishes) unless there is an additional frequency shift for the central transition as well, due for example to a chemical shift non-equivalence.

If second-order quadrupolar effects cannot be neglected, $\bar{\mathscr{H}}_{\mathrm{Q}}^{(2)}$ has to be considered as well, and analytical expressions for $\bar{\mathscr{H}}_{\mathrm{Q}}^{(2)}$ are rather lengthy. However, when considering a given transition, $\bar{\mathscr{H}}_{\mathrm{Q}}^{(2)}$ can be viewed as a fictitious spin-1/2 operator, in which case we have $\bar{\mathscr{H}}_{\mathrm{Q}}^{(2)} = \Omega s_z$, where Ω is the orientation dependent second-order frequency shift, and s_z denotes the z-component of the spin-angular momentum operator for $I = 1/2$. Since a selective π-pulse converts s_z into $-s_z$, for a Hahn echo sequence, the second-order quadrupolar interaction is greatly reduced, as in the case of a resonance offset effect. However, as long as the Zeeman levels represent stable states within the time scale of the formation of the spin-echo, the influence of the second-order quadrupolar interactions can be neglected when using a selective $\pi/2 - \pi$ pulse sequence. This will be the subject of a forthcoming paper by one of us (J. H.).

The effect of a 90° phase-shift between the $\pi/2$ and the π pulse, as long as a selective excitation is guaranteed, will not change the spin-echo decay, except that the echo is now in-phase with the one-pulse free induction decay.

For a selective $(\pi/2)_x - (\pi/2)_y$ pulse sequence similar numbers for $E_{\mathrm{L}}^{(0)}$, within about 10%, have been reported [75]. However, since the second pulse is now a $\pi/2$ pulse, it does not reduce the second-order quadrupolar interaction as is the case with a π pulse. Also, resonance offset effects are not eliminated. Especially for powders, with a spread in second-order frequency shifts, or a chemical shift anisotropy, a faster echo decay due to these resonance offset effects can be observed. This effect can even dominate with slowly decaying echoes, as found, for example, with several zeolites having low Al-content [75].

If homonuclear and heteronuclear dipolar interactions are present the calculation of the second moment becomes rather complicated. However, the effect of an additional heteronuclear dipolar coupling can be approximated by means of *reduced second moments* [75]. As was shown by Kambe and Ollom [73] and by Mansfield [74], the frequency shift due to the first-order quadrupolar interaction restricts spin-exchange to the same transition of neighboring nuclei. This is why, for calculation of the second moment of the central transition, only the matrix elements of the homonuclear dipolar interaction (which are diagonal or describe the spin-flipping between the same transitions) have to be added. We have also seen for a $\pi/2 - \pi$ pulse sequence that the spin-echo decay of the central transition is overwhelmingly dominated by the homonuclear dipolar interactions of the spins populating the Zeeman levels with $m = \pm 1/2$. Therefore, it should be possible to describe the spin-echo decay of the central transition by its ordinary second moment for the one-pulse response, where we discard for the sum of the dipolar matrix elements all those elements which involve Zeeman levels for which $|m| > 1/2$. One thereby calculates a *reduced* second moment, characterized by the following spin-dependent factor [75], cf. Eqs. (3.17–3.20). For the ordinary second moment is

$$F_{\mathrm{ex}}^{\pm 1/2} = \frac{1}{9}\left\{\frac{2}{(2I+1)} + \frac{(2I+1)}{2} + \frac{(2I+1)^3}{32}\right\}, \tag{5.35}$$

where the labels "ex" and "$\pm 1/2$" refer to the fact that only the central levels are excited and spin-exchange among the central levels is still allowed. Comparison with $E^{(0)}/F^{(0)} \approx 0.5$ shows that the reduced second moment for the *free induction decay* describes very well the $\pi/2 - \pi$ spin-echo decay envelope for the selectively-excited central transition.

The resonance frequencies of the *central* transitions of two neighboring I-spins can be different. This might be caused by very different quadrupolar couplings, or by chemical shift effects. In this case the spin-exchange between the central levels of neighboring I-spins is completely suppressed. In order to describe the change in the spin-echo decay one again calculates a virtual (reduced) second moment of the free induction decay. Summing up only matrix elements of the homonuclear dipolar interaction involving the central levels, $m = \pm 1/2$, but, discarding also all matrix elements responsible for the spin-exchange. One finds that for spins with $m = \pm 1/2$ only the diagonal part of the homonuclear dipolar interaction contributes to the second moment. For the reduced second moments, the following spin-dependent factor has been calculated [75]:

$$F_{\mathrm{nex}}^{\pm 1/2} = \frac{2}{9(2I+1)}. \tag{5.36}$$

Comparison of Eq. (5.36) with Eq. (5.28) shows that the total suppression of spin-exchange among the central levels causes a large decrease in the corresponding second moment. These results can now be used to describe the influence of an additional heteronuclear dipolar interaction on the spin-echo decay of the central transition.

By means of these reduced second moments it is possible to find expressions for the spin-echo decay if homonuclear and heteronuclear dipolar interactions are present. The non-resonant S-spin system is responsible for two main effects. First, by influencing the resonance frequency of the I-spin (IS-interaction), which causes line broadening, it suppresses the spin-flipping between like nuclei, and thus the spin-echo decay is slowed down. Second, spin-flipping between the S-spins (SS-interaction) will decrease the phase coherence in the I-spin system, which tends to destroy the formation of the spin-echo (as for a considerably shorter spin–lattice relaxation of the S-spins). In order to describe the spin-echo decay for two coupled spin systems quantitatively, one makes the assumption that the above mentioned processes can be approximated by the corresponding second moments: M_{2F}^{II}, M_{2F}^{IS}, M_{2F}^{SS}, M_{2F}^{SI}, higher-order moments being neglected. As mentioned above, the spin-echo decay for a purely homonuclear dipolar interaction is well described by the reduced second moment, where one takes only the central levels into account. Upon introducing an additional heteronuclear coupling, the spin-flipping in the I-system vanishes completely, since the secular part of the homonuclear interaction no longer contains x and y components of the angular momentum operators (which leads to further truncation of $\mathscr{H}_D$). Thus, the second moment of the spin-echo decay for the I spin system, denoted as $M_{2E}(I)$, in the presence of a heteronuclear dipolar coupling decreases drastically, to the value

$$M_{2E}(I) = M_{2F}^{II} \frac{F_{nex}^{\pm 1/2}(I)}{F_L(I)}. \tag{5.37}$$

If a heteronuclear coupling causes this drastic decrease, the coupling to the S-spin system, M_{2F}^{IS}, will in general be larger than $M_{2E}(I)$. So, the two spin systems appear to be strongly coupled, and the influence of the S-spin system can be approximated by assuming an echo intensity, $E(2\Delta)$, having the following form:

$$E(2\Delta) = \exp\left\{-\frac{M_{2E}(I)}{2}(2\Delta)^2\right\}\exp\left\{-\frac{M_{2E}(S)}{2}(2\Delta)^2\right\}, \tag{5.38}$$

where $M_{2E}(I)$ refers to the second moment of the spin-echo decay, Eq. (5.37), in the I-system, if no spin-flipping occured in the S-spin system.

If both I and S-spin systems are subjected to strong quadrupolar coupling, $M_{2E}(S)$ has to be calculated from Eq. (5.37). In the special case that the S-spins have a vanishing quadrupolar coupling (spin-exchange is not restricted), or $S = 1/2$, $M_{2E}(S)$ is given by van Vleck's formula for the S-spin homonuclear dipolar interaction, Eq. (3.17).

The $1/e$ decay time deduced from Eq. (5.38) for a Gaussian decay of the spin-echo now becomes

$$T_{2E} = \sqrt{\frac{2}{M_{2E}(I) + M_{2E}(S)}}. \tag{5.39}$$

If the heteronuclear dipolar coupling is negligible, $M_{2E}(S) = 0$ in Eq. (5.39), then $M_{2E}(I)$ can be replaced by M_{2E}, and one obtains

$$T_{2E} = \sqrt{\frac{2}{M_{2E}}}, \tag{5.40}$$

which is the ordinary relation between the $1/e$ decay time and the second moment for a Gaussian decay.

6 Sample Rotation

6.1 Spinning About One Axis

The magic-angle spinning (MAS) technique has mainly been applied to spin-1/2 nuclei [46, 128–131]. The behavior of the several spin interactions under MAS is discussed by Mariqc and Waugh [132]. Although, the "magic" angle should be defined with respect to the considered interaction [45], as widely common, we use the definition arising from $(3\cos^2\theta_m - 1) = 0$, where θ_m stands for the magic angle.

The first-order quadrupole interaction can be viewed as an inhomogeneous interaction [133] and each satellite transition behaves exactly the same as a chemical shift anisotropy. The central transition has no orientation dependence in first-order and is thus unaffected by MAS. Spinning sidebands appear at frequencies $\omega_L + n\omega_{rot}$, where $|n| = 1, 2, 3, \ldots$ The line at ω_L is usually called "center band". Hence, for quadrupole nuclei the center band is the sum of the $2I - 1$ center bands of the satellite transitions and the central transition. The center bands due to the satellite transitions must not be confused with the central transition. If the central transition is broadened by other interactions, it may also split into center band and various spinning sidebands. In the time-domain the satellite transitions appear as rotational echoes at times, $t = n/\nu_{rot}$. If there is a total of N spinning sidebands in the frequency domain, the echoes in the time-domain are a superposition of N cosine functions, $\cos(n\omega_{rot}t)$. The envelope of the echo train is given by the inverse linewidth of a single MAS line. The magnitude of such a single echo is proportional to the total number of sidebands, N, and its width proportional to $1/N$. Therefore, by maximizing the magnitude of an echo, a maximum number of spinning sidebands will be observed after the Fourier transform. This shows that the first-order quadrupole interaction may be used to set the magic angle [134].

Since the first-order quadrupole shift is typically bigger than the MAS rotation frequency, the envelope of the first-order MAS powder pattern can be used to determine the quadrupole parameters. However, very often the whole spectrum cannot be excited and therefore, only few studies of nuclei with half-integer spin report the use of a high-performance broad band MAS probe for the determination of quadrupole parameters (up to $C_{qcc} = 2.5$ MHz) from the first-order powder sideband envelope [135–138].

Although the second-order quadrupole shift is not completely averaged by using MAS, working at the magic angle still has the advantage of removing effects due to dipole interactions and chemical shift anisotropy, making line-shape simulations of the central transition more accurate compared with static powder spectra. Variable angle spinning (VAS) and newer techniques such as dynamic angle spinning (DAS) and double angle rotation (DOR) with potential application for second-order broadened lineshapes will also be discussed in this chapter.

Nolle [139] first reported the narrowing of the central line of ^{95}Mo in Mo(CO)$_6$, upon sample rotation about an axis perpendicular to the static magnetic field and calculated the theoretical powder spectra for $\eta = 0$. The first paper concerning MAS studies of half-integer nuclei were published in 1981 [140–142]. Oldfield and co-workers [143–145] established variable-angle-spinning (VAS) NMR technique and presented computer simulated spectra as a function of the angle and for various values of η. Behrens and Schnabel [72] calculated the VAS lineshape in the case of $\eta = 0$ and the MAS lineshape for various values of η. Similar results were obtained by Müller [69], who used the approach of Narita et al. [56] for the calculation of the second-order resonance shift of the central line, cf. Sect. 3, and calculated the narrowing factor in dependence on θ for $\eta = 0$, as shown in Fig. 6.1, cf. Ref. [146]. A wealth of MAS and VAS lineshape calculations including the spinning sidebands can be found in the literature since 1982. Three recent studies should be mentioned first: Zheng et al. [43] extended the powder average approach of Alderman et al. [66] and a method introduced by Sethi et al. [67] for the computation of sideband intensities to a simple formalism for computing NMR spectra of the central transition of quadrupole nuclei with half-integer spin in powdered solids spinning at any angle and at any speed. Amoureux et al. [147, 148] considered the rf pulse effects for the lineshape calculations. Skibsted et al. [65] introduced a method which allows the determination of quadrupole and anisotropic shielding tensors, including the relative orientation of their principal-axis system from first-order MAS sideband spectra.

A comprehensive theoretical investigation of the VAS NMR lineshape of quadrupole nuclei was presented by Lefebvre et al. [149]. The authors neglected dipolar interactions and the anisotropy of the chemical shift, and concluded that there are two favorable angles of rotation: $\theta = 43.5°$ allows an easy determination of ν_Q and, η, and, $\theta = 60°$ to $70°$ gives the maximum line narrowing. But, the VAS NMR technique has found only very few experimental applications for systems which have relatively "diluted" quadrupole nuclei: ^{23}Na, ^{51}V, and ^{55}Mn in

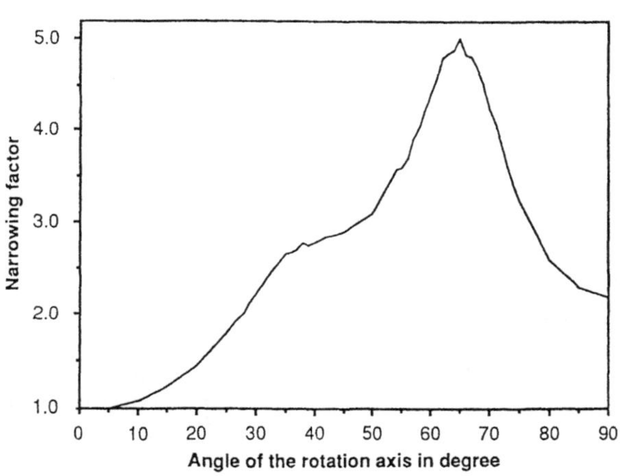

Fig. 6.1. Narrowing factor $\sqrt{M_2/M_2^{rot}}$ for the second-order quadrupole broadened center band of the central transition in the case of $\eta = 0$ and fast rotation about axes inclined at the given angle with respect to the direction of the external magnetic field. The values of the second moments were numerically calculated by D. Müller with respect to the center of gravity, cf. Ref. [146]

several inorganic solids [43, 143, 150], ^{11}B in borosilicate glass [145, 150], and ^{27}Al in Zunyite [151].

Spinning sidebands of VAS spectra have been simulated [150, 152] and compared with experimental spectra. If the rotation frequency exceeds the static width of the central transition, more than 95% of the total central transition intensity appears in the center band [152], satellite transitions excluded, and this is independent on the angle of rotation, θ. However, for slower rotation, the relative intensity of the center band becomes dependent on θ and η. Amoureux et al. [152] showed that the relative intensity, X, is a function of

$$X = \frac{\nu_Q^2}{4\nu_L \nu_{rot}} [I(I+1) - 3/4]\left(1 + \frac{\eta^2}{3}\right) = \frac{4A}{\nu_{rot}}\left(1 + \frac{\eta^2}{3}\right) = \frac{\Delta}{\nu_{rot}}. \qquad (6.01)$$

Since the spectral width of the central transition without sample rotation is $A(25/9 + 14\eta/9 + \eta^2/9)$, cf. Sect. 3, the value $4A(1 + \eta^2/3) = \Delta$ is comparable with the static spectral width. With respect to θ, the maximum intensity and a relatively weak dependence on η is obtained for $\theta = 90°$, e.g. 80% for $\nu_{rot} = \frac{\Delta}{5}$ [152]. If the rotation axis is at the magic angle ($\theta_m = \arccos(1/\sqrt{3})$, for $\nu_{rot} \geqq \Delta$, the sum of sideband-intensities is less than 5% of the total intensity of the center band. This value increases to 46% if the rotation frequency decreases to $\nu_{rot} = \frac{\Delta}{5}$ and $\eta = 0$ [138, 152]. Thus, if the rotation frequency is smaller than the spectral width of the central transition without rotation, either the next order sidebands must be added to the center band or the amount of the central transition intensity in the center band must be calculated for intensity measurements. Figure 6.2, taken from Ref. [138], shows the amount of intensity of the central transition in the center band as a function of $\nu_Q^2/\nu_L\nu_{rot}$ for $I = 5/2$. In addition, for $\nu_{rot} \approx \Delta/4$, a superposition of the center band and the next spinning sidebands reduces the resolution of the bands and requires a fitting with calculated spectra. The intensity of spinning sidebands due to the satellite transitions increases with the

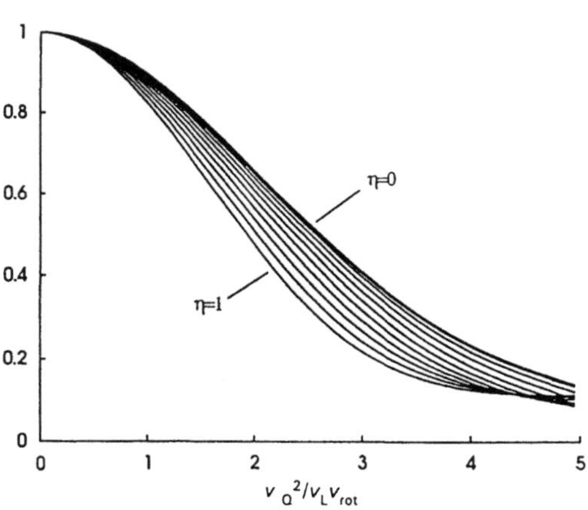

Fig. 6.2. Amount of intensity of the central transition in the center band for $I = 5/2$ toward $\nu_Q^2/\nu_L\nu_{rot}$ taken from Massiot et al. [138]

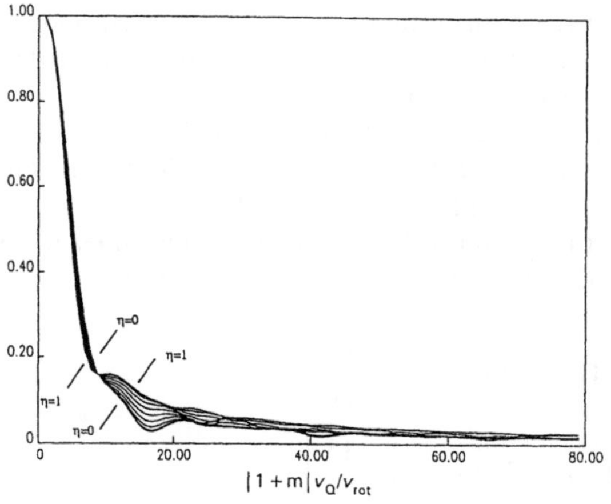

Fig. 6.3. Amount of intensity of the satellites (due to the $m \rightarrow m+1$ transitions) in the center band as a function of $|1+m|\nu_Q/\nu_{rot}$ taken from Massiot et al. [138]

rotation frequency, which is small compared with the first-order spectral width, whereas for the central transition for $\nu_{rot} > \Delta$ essentially all the intensity appears in the center band. Since the intensity of the first-order spinning sidebands is quite constant near the center band, the contribution of the satellites to the center band can be estimated from sidebands near the center band, containing no contributions from the central transition. For strong quadrupole coupling, if the quadrupole frequency exceeds the rotation frequency by more than two orders of magnitude, this contribution can be neglected. Figure 6.3 shows the intensity of the satellites in the center band as a function of $|1+m|\nu_Q/\nu_{rot}$ [138].

Samoson [153] has shown that the second-order quadrupole shift of the $m \leftrightarrow m+1$ transition can be split into an isotropic part, ν_{cg}, describing the center of gravity and an angular dependent part $\nu(\alpha, \beta, \eta)$:

$$\nu^{(2)}_{m;m+1} = \nu_{cg} + \nu(\alpha,\beta,\eta) = -\frac{\nu_Q^2}{30\nu_L}[I(I+1) - 9m(m+1) - 3]\left(1 + \frac{\eta^2}{3}\right)$$
$$-\frac{\nu_Q^2}{30\nu_L}\left[I(I+1) - \frac{17}{3}m(m+1) - \frac{13}{6}\right]g(\alpha,\beta,\eta), \tag{6.02}$$

where the orientation-dependent function g can be described by $g(\alpha,\beta,\eta) = A(\alpha,\eta)\cos^4\beta + B(\alpha,\eta)\cos^2\beta + C(\alpha,\eta)$ with

$$A = \frac{105}{16} - \frac{35}{8}\eta\cos 2\alpha + \frac{35}{48}\eta^2\cos^2 2\alpha,$$

$$B = -\frac{45}{8} + \frac{5}{12}\eta^2 + 5\eta\cos 2\alpha - \frac{35}{24}\eta^2\cos^2 2\alpha,$$

$$C = +\frac{9}{16} + \frac{1}{3}\eta^2 - \frac{5}{8}\eta\cos 2\alpha + \frac{35}{48}\eta^2\cos^2 2\alpha.$$

It should be mentioned, again, that in our notation $m = -1/2$ corresponds to the central transition, in other notations $m = +1/2$ represents the central transition [142, 153], cf. remarks below Eq. (2.19).

For the central transition, $m = -1/2$, Eq. (6.02) corresponds to Eq. (3.05) with (3.07) and the isotropic part in Eq. (6.02) corresponds to Eq. (3.10). The function $g(\alpha, \beta, \eta)$ as given in Ref. [153] contains some mistakes.

The first equation describing the second-order quadrupole shift under MAS condition was derived by Kundla et al. [142]. It reads in our notation

$$\begin{aligned} \nu^{(2)}_{m;m+1} = &-\frac{\nu_Q^2}{12\nu_L}\left[I(I+1) - 7m(m+1) - \frac{5}{2}\right]\left(1 + \frac{\eta^2}{3}\right) \\ &+ \frac{\nu_Q^2}{12\nu_L}\left[3I(I+1) - 17m(m+1) - \frac{13}{2}\right] \\ &\cdot\left\{\sin^2\beta\left(1 - \frac{7}{8}\sin^2\beta\right) - \frac{\eta}{6}\sin^2\beta\left(3 - \frac{7}{2}\sin^2\beta\right)\cos 2\alpha\right. \\ &\left. + \frac{\eta^2}{18}\left[1 + \sin^2\beta\left(1 - \frac{7}{4}\sin^2\beta\cos^2 2\alpha\right)\right]\right\}. \end{aligned} \tag{6.03}$$

The angular-independent term in Eq. (6.03) does not represent the center of gravity. For $m = -1/2$, Eq. (6.03) gives Eq. (3.05) with (3.07). It should be noted that Eq. (6.03) corresponds to the original equation which is given in another form in Ref. [142]. The angular-dependent term is misprinted in Refs. [154, 155].

Using Eq. (6.02), the true value of the chemical shift (which is included in the Larmor frequency in this equation) can be determined [135, 153, 156–158]. For nonselective excitation, or at least partially selective excitation, sidebands can be observed which are just outside of the spectral range of the static spectrum of the central transition. Their main intensity results from the $\pm\frac{1}{2}\leftrightarrow\pm\frac{3}{2}$ transitions, whereas the center band results mainly from the $+\frac{1}{2}\leftrightarrow-\frac{1}{2}$ transition. The average resonance position of two equal-order-sidebands $\frac{1}{2}(\nu_{i,cg} - \nu_{-i,cg})$ can be taken as the center of gravity of the $\pm\frac{1}{2}\leftrightarrow\pm\frac{3}{2}$ transitions, $\nu_{cg\pm 3/2,\pm 1/2}$.

The difference $\Delta = \nu_{cg\pm 1/2,\pm 3/2} - \nu_{cg-1/2,+1/2}$ to the center of gravity of the center band results from Eq. (6.02):

$$\Delta = -\frac{1}{30}\frac{\nu_Q^2}{\nu_L}9\left(1 + \frac{\eta^2}{3}\right). \tag{6.04}$$

The combination of Eqs. (6.02) and (6.04) gives the quadrupole shift of the center band, which is for $I = \frac{5}{2}$

$$\nu_{cg-1/2,+1/2} - \nu_L = \frac{8}{9}\Delta. \tag{6.05}$$

This procedure for the determination of the quadrupole shift or other quadrupole parameters was introduced by Samoson [153]. It is convenient for $I=\frac{5}{2}$ nuclei, because the linewidth of their $\pm\frac{3}{2}\leftrightarrow\pm\frac{1}{2}$ satellites is decreased by a factor of 0.3 with respect to the central line. The procedure is less efficient for other nuclei or transitions [153]. Jäger et al. used this sideband analysis for the study of ^{27}Al nuclei in various inorganic compounds [158–160, 297] and denoted it as SATRAS (satellite-transition spectroscopy). If the whole spectrum can be excited, the splitting of spinning sidebands in a distance of ca. $0.6\nu_Q$ to the center band, which consist of similar intensities arising from the $\pm\frac{5}{2}\leftrightarrow\pm\frac{3}{2}$ and the $\pm\frac{3}{2}\leftrightarrow\pm\frac{1}{2}$ transitions gives also the information about the quadrupole shift [135].

TOSS, a method for total suppression of spinning sidebands, which is widely used in solid-state ^{13}C MAS NMR [161] can improve the resolution of ^{27}Al MAS NMR spectra, as was shown by Carduner [162], if the rotation frequency is comparable with the MAS linewidth. The reported low signal-to-noise is likely to be caused by partly selective pulses and could have been improved by tuning to the central transition, only.

Applications of MAS NMR represent the majority of NMR investigations of half-integer spin quadrupole nuclei during the last decade. MAS NMR is an important method for studies of inorganic solids and minerals, especially for zeolites and other highly porous materials [129, 130, 155, 163, 164], glasses and ceramics [165, 166], and for group III–V semiconductors [167]. During the last few years the applications have benefited from higher magnetic fields, up to 600 MHz proton resonance frequency in combination with high-speed MAS up to 20 kHz [168–170]. Under these conditions, e.g., ^{27}Al nuclei having quadrupole coupling constants up to about 15.5 MHz have been investigated by MAS NMR [171].

6.2 *MAS With Cross-Polarization*

Cross-polarization (CP) was introduced by Hartmann and Hahn [172] and applied to the proton enhanced NMR of dilute spins in solids by Pines, Gibby and Waugh [173]. A detailed discussion of the CP experiment can be found by Slichter [174]. Despite of some difficulties at high spinning frequencies, the combination of CP with sample rotation is widely used now for solid-state studies concerning low-sensitivity nuclei. S. Vega [29] discussed static CP of half integer quadrupole nuclei and showed some results on single crystals.

The polarization transfer from ^{1}H to quadrupole nuclei of half-integer spin combined with MAS has found some applications: ^{11}B CP MAS NMR by Woessner [175], ^{17}O CP MAS NMR by Walter et al. [176, 177], ^{23}Na CP MAS NMR by Harris and Nesbitt [178], ^{27}Al CP MAS NMR by Blackwell and Patton [179], Morris et al. [180, 181], Kellberg et al. [182], Rocha and Klinowski [183–185], ^{43}Ca CP MAS NMR by Bryant et al. [186], and ^{95}Mo CP MAS NMR by Edwards and Ellis [187, 188].

The above mentioned applications were limited to ^1H and the central transition of the quadrupole nuclei with spin S. In this case, the Hartmann–Hahn equation corresponds to the nutation frequencies of selective excitation, cf. Sect. 2,

$$(S + 1/2)\gamma_S B_{rf\,Q} = \gamma_H B_{rf\,H}. \tag{6.06}$$

A. J. Vega [189, 190] considered the spin dynamic of CP MAS NMR of spin-3/2 nuclei in dependence on the ratio $\alpha = \nu_{rf}^2/\nu_Q\nu_{rot}$. He found a greatly reduced spin-lock efficiency for the intermediate case $\alpha = 0.4$ and more efficient polarization transfer for $\alpha \gg 1$ and $\alpha \ll 1$. However an upper limit of α is given by the overlap of sidebands for slow spinning speeds. In the limit of fast spinning, the Hartmann–Hahn matching must satisfy a sideband condition.

$$\nu_{rf\,H} = (S + 1/2)\nu_{rf\,S} \pm n\nu_{rot} \tag{6.07}$$

with $n = 1, 2$. The important result of these studies [189, 190] is that for strong quadrupole coupling the fastest possible spinning speed is recommended.

Cross polarization transfer is usually benefitted from the ratio of the heat capacities of the I spins and S spins in the rotating frame. If the energy exchange is fast compared with relaxation effects, the S spins can be cooled down to the temperature of the I spins, thus, the magnetization of the S spins can be increased by a factor of γ_I/γ_S, cf. also Mehring [46]. Morris et al. [181] claimed on the basis of a temperature concept that the above mentioned gain in magnetization of the S spins decreases to $\gamma_I/(S + 1/2)\gamma_S$, if the polarization transfer takes place between a spin-1/2 system (high heat capacity) with a effective spin-1/2 system of quadrupole nuclei (low heat capacity). Conversely, Walter et al. [176] found a gain near to γ_I/γ_S for an ^1H-^{17}O CP MAS experiment.

In conclusion, the selectivity of the CP MAS experiment has been shown to be helpful for distinguishing quadrupole nuclei which have dipolare coupling to protons. A comprehensive understanding of the CP process for quadrupole nuclei as a function of the dipole interaction is not yet known.

6.3 Dynamic Angle Spinning and Double-Rotation

The quadrupolar broadening of the central transition for powders can be reduced using MAS or VAS. Since the narrowing factor is less than 10 (ca. 3.6 at the magic angle for fast rotation), the second-order line width remains the major obstacle for high resolution. With ν_{cg} as the center of gravity of the lineshape, the second-order quadrupole shift can be expressed by its angular dependent contribution $A_\ell(\phi, \psi)$, where ϕ, ψ are Euler angles describing the transformation between the quadrupole PAS and the rotor axis

$$\nu - \nu_L = \nu_{cg} + \sum_{\ell = 2;4} A_\ell(\phi, \psi) P_\ell(\cos\beta). \tag{6.08}$$

The terms $P_\ell(\cos\beta)$ depend on the rotor axis orientation with respect to the external field, Euler angle β, and are given by the second-order and forth-order Legendre polynominals

$$P_2(\cos\beta) = \tfrac{1}{2}(3\cos^2\beta - 1), \quad P_4(\cos\beta) = \tfrac{1}{8}(35\cos^4\beta - 30\cos^2\beta + 3). \quad (6.09)$$

For $P_2 = 0$ it is inferred that $\beta = \arccos(1/\sqrt{3}) \approx 54.74°$, which is the magic angle. $P_4 = 0$ yields $\beta = \arccos(\sqrt{\tfrac{3}{7} \pm \tfrac{2}{7}\sqrt{\tfrac{6}{5}}}) \approx 30.55°$ or $70.12°$. Based on these considerations new experimental techniques, DAS and DOR, have been developed, recently, cf. Llor and Virlet [191], Samoson et al. [192], and Chmelka et al. [193]. A comprehensive description of the theory of DAS and DOR NMR including the sideband analysis was presented by Sun et al. [298].

DAS, the dynamic-angle spinning technique, indroduced by Pines and co-workers [193–195] and Llor and Virlet [191], is based on the sample spinning about an axis inclined at a time dependent angle $\beta(t)$ with respect to the external field. In the most simple case $\beta(t)$ toggles between the two fixed angles oft 37.38°, and 79.19° for two equal periods of time. During the time, necessary for switching the angle (ca. 30 ms), it is necessary to minimize the evolution of the magnetization which can be achieved by means of $\pi/2$ pulse pairs. The pulses must be selective and calibrated corresponding to the angle of rotation. The signal being a function of the time of evolution, t_1($t_1/2$ for each angle of rotation), yields a second-order quadrupole echo at $t = t_1$, after the second pulse. The envelope of the echoes as a function of t_1 gives a 1D spectrum, or, by an additional data acquisition of the FID starting at the echo maximum, a 2D spectrum can be obtained.

Advanced concepts of the DAS technique were presented by Mueller et al. [195] who introduced a fourth $\pi/2$ pulse in order to retain two pathways from the t_1 evolution, yielding a pure absorption-phase spectrum. A third angle of rotation, $\beta = 54.74°$, has been introduced [195] in order to correlate in two dimensions the DAS spectra with MAS spectra. Sideband pattern emerges if the evolution period of the 2D DAS experiment is divided into unequal parts by rf pulses or spinner reorientation: Grandinetti et al. [196] obtained ^{87}Rb spectra over a range of DAS angles and described a theory of the sideband pattern. They found that the angle pair (0°, 63.43°) gives sidebands at integer multiples of 0.83 ω_{rot} and should be particulary useful for DAS cross-polarization experiments involving central transitions of quadrupole nuclei [196].

There are two drastic limitations for the application of the DAS-technique. First, the spin–lattice relaxation time, T_1, often short for quadrupole nuclei, has to be larger than the time period necessary for flipping the rotor axis. Second, spin exchange due to dipolar interactions cannot be eliminated during the relatively long flipping time of the rotor axis, and has to be sufficiently small.

DOR, the double-rotation technique, was introduced by Samoson, Lippmaa, and Pines [192] who succeeded in 1988 in building a double-rotor probe. The outer rotor, inclined by $\beta = 54.74°$ with respect to the external field, rotated at about 400 Hz, and had a diameter of 20 mm. The inner rotor, with a diameter of 5 mm, rotated at about 2 kHz, and the angle between both axis of rotation is $\beta_1 = 30.6°$

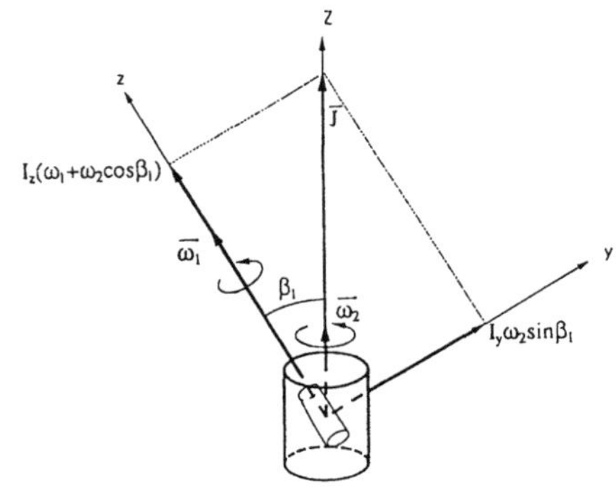

Fig. 6.4. Picture of the moments of inertia for double rotation, taken from Wu et al. [199]. By adjusting the ratio of moments, I_y/I_z, the addition of the two vector components of the angular momentum along the inner-rotor axis z and the y-axis can be made to point along the outer-rotor axis Z, so that spinning the outer rotor will not affect the orientation of the total angular momentum of the inner rotor

[197, 198]. The rotation must be torque-free. This can be achieved only for certain ratios of ν_{inner}/ν_{outer}. The two contributions to the inner-rotor angular velocity are ω_1 and the angular velocity ω_2 imposed by the outer rotor. The later can be decomposed into one component $\omega_2 \cos\beta_1$ along the z-axis of the inner rotor and another component $\omega_2 \sin\beta_1$ perpendicular to this axis in y-direction, so that y-direction, axis z of the inner rotor, and axis Z of the outer rotor are in one plane. Thus, $\omega_x = 0$, $\omega_y = \omega_2 \sin\beta_1$, $\omega_z = \omega_1 + \omega_2 \cos\beta_1$, and the resultant vector $\boldsymbol{J}$ must be in Z-direction in order to be torque-free, cf. Fig. 6.4, taken from Ref. [199]:

$$\tan\beta_1 = \frac{I_y \omega_2 \sin\beta_1}{I_z(\omega_1 + \omega_2 \cos\beta_1)}, \tag{6.10}$$

or

$$\frac{\omega_1}{\omega_2} = \cos\beta_1 \left(\frac{I_y}{I_z} - 1\right) \equiv k. \tag{6.11}$$

Wu et al. [199] proved that the frequencies obey the Equation

$$\nu_{inner} = k\nu_{outer} + \Delta\nu_{inner}, \tag{6.12}$$

where $k = 0.49$–5.4 and the excess of the inner frequency is $\Delta\nu_{inner} = 0.6$–1.4 kHz.

Compared with the DAS technique, DOR does not suffer from the loss of coherences during the flip time. The application of rotation-synchronized pulses has been shown to be effective in eliminating half of the DOR sidebands [197]. The DOR technique has been applied, at first, to the ^{23}Na resonance in sodium oxalate [192, 197, 198], $NaNO_2$ [197], and ^{17}O in diopside [193]. With a new double-rotor design [199] structural changes in porous aluminium phosphates have been studied by ^{27}Al DOR NMR [200–204] and silicates have been studied by ^{17}O DOR NMR [205]. Müller et al. [206] measured with the DOR technique the ^{23}Na signal of $NaNO_2$ and the ^{27}Al signal of YAG. Jelinek et al. studied a sodalite semiconductor supralattice [207] and compounds encapsulated in zeolite NaY [208]. The number of DOR NMR studies is drastically increasing now. Also cross-polarization can be included [299].

DOR NMR gives the most accurate values of the (observed) shift, δ_{obs}, of the nuclear magnetic resonance of a quadrupole nucleus. The isotropic quadrupole shift δ_Q must be determined by additional methods in order to obtain the value of the chemical shift $\delta_{CS} = \delta_{obs} - \delta_Q$. The application of DOR NMR at different magnetic field strengths yields also the value of δ_Q, and sites with equal values of δ_{CS} but different electric field gradients can be discriminated at lower field strength [201, 203].

The rotation frequencies of the inner and outer rotor presently do not exceed $\nu_{inner} = 5$ kHz and $\nu_{outer} = 1$ kHz, respectively. The rotation frequency of the inner rotor limits the resolution of the DOR spectra: ν_{inner} must be greater than the linewidth arising from the homonuclear dipole–dipole interaction [200]. The residual linewidth of a MAS spectrum of a first-order quadrupole broadened spectrum does not depend on the first-order quadrupole interaction. The same can be expected for DOR and second-order quadrupole broadening. However, a distribution of the quadrupole parameters causes a distribution of isotropic quadrupole shifts, which broadens the DOR signal.

In conclusion, the application of the DOR technique is in progress now. But fast MAS in high magnetic fields is still the most important experimental technique for lineshape and chemical shift analysis of quadrupole nuclei in solids. It is easy to use and the experimental setup is more straightforward than for (VAS) DAS and DOR. Measurements in a wide temperature range can be performed.

6.4 *Magic-Angle Spinning Spectra of Spin-1/2 Nuclei Coupled to Quadrupole Nuclei*

The title of the given review "Quadrupole Effects in Solid-State NMR" involves the well-known effect that MAS fails to completely eliminate the effect of dipolar coupling for spin-1/2 nuclei when coupled to quadrupole nuclei with a quadrupole frequency comparable to the Zeeman frequency of the quadrupole nuclei. This phenomenon was first discovered for the ^{13}C–^{14}N pair [209] and the majority of studies has been devoted to spin-1/2-spin-1 pairs (c.f. [210]). Because this review deals with half-integer spin quadrupole nuclei, mainly the spin-1/2 – spin > 1 pairs will be considered, for which less attention has been paid: ^{31}P–^{63}Cu (Menger and Veeman [211], see also Olivieri [212]); ^{31}P–^{27}Al (Chu et al. [213]); ^{31}P–^{55}Mn, ^{59}Co, ^{93}Nb (Gobetto et al. [214]); ^{13}C–^{23}Na, ^{29}Si–^{35}Cl (Böhm et al. [215]); ^{119}Sn–$^{35,37}Cl$ (Komoroski et al. [216], Harris [217], Apperley et al. [218]); ^{13}C–^{75}As (Sastry et al. [219]); ^{31}P–^{63}Cu (Olivieri [220]).

Olivieri [212, 221] presented an analytical description for the centers of gravity of the single lines for any I–S pair. The scalar I–S coupling gives rise to the usual multiplet structure, but also to a shift, analogous to that discussed for the dipolar interaction with the quadrupole nuclei. Apperley et al. [218] discussed a system, where a spin-1/2 nucleus is coupled to two equivalent spin-3/2 nuclei, see also Ref. [213, 214]. A review about "Quadrupolar effects transferred to spin-1/2 MAS spectra of solids" was presented by Harris and Olivieri [300].

Olivieri [212] estimated the narrowing effect by MAS and calculated ν_m^{iso}, the isotropic shift of the line ν_m of one resonant spin-1/2, dipolar coupled to one quadrupole nucleus in the state m (first-order perturbative approach, spatial averaging for powder samples, and fast sample spinning so that spinning sidebands do not overlap with the center band):

$$\nu_m^{iso} = \frac{\nu_{IS}\nu_Q}{10\nu_S}[S(S+1) - 3m^2](3\cos^2\beta - 1 + \eta\sin^2\beta\cos 2\alpha). \tag{6.13}$$

The dipolar splitting, ν_{IS}, is the frequency distance between the singularities of the inner Pake's doublet of the static powder spectrum of an I–S spin pair of distance r_{ik}:

$$\nu_{IS} = \frac{\mu_0}{4\pi}\frac{\gamma_I\gamma_S h}{4\pi^2}\frac{1}{r_{ik}^3}. \tag{6.14}$$

The quadrupole frequency, cf. Eq. (1.16), is

$$\nu_Q = \frac{3\,e^2qQ}{2S(2S-1)h}. \tag{6.15}$$

The Zeeman frequency of the non-resonant quadrupole nucleus is ν_S, η is the asymmetry parameter, and the angles β and α define the orientation of the vector connecting the spin pair in the PAS of the quadrupole interaction, cf. Sect. 1.3. The shift can be positive or negative, or even disappear for the hypothetical case $\beta = \arccos(1/\sqrt{3})$, and $\eta = 0$. The geometry factor in Eq. (6.13) is 2 if the dipolar and quadrupolar principle axis systems are coincident ($\beta = 0$).

An estimate of the line narrowing by MAS, is given by the narrowing factor $(M_2^{VanVleck}/M_2^{MAS})^{1/2}$. Van Vleck's second moment (powder) is for a heteronuclear spin pair (in squared frequency units), cf. Eqs. (6.14) and (3.18),

$$M_2^{VanVleck} = \frac{4}{15}\nu_{IS}^2 S(S+1). \tag{6.16}$$

If spinning sidebands are not considered and ν_Q is extremely small, M_2^{MAS} vanishes. For a stronger quadrupole interaction M_2^{MAS} can be calculated summing up the values of ν_m^2, given by Eq. (6.13):

$$M_2^{MAS\,iso} = \frac{1}{2S+1}\sum_{m=-S}^{m=+S}\nu_m^2 = M_2^{VanVleck}\frac{\nu_Q^2}{\nu_S^2}(2S-1)(2S+3)\frac{3}{400}$$
$$\cdot(3\cos^2\beta - 1 + \eta\sin^2\beta\cos 2\alpha)^2 \tag{6.17}$$

Fenzke and Brunner [222] calculated both, the isotropic and the anisotropic contribution to the second moment and obtained

$$M_2^{MAS\,iso+aniso} = M_2^{VanVleck}\frac{\nu_Q^2}{\nu_S^2}(2S-1)(2S+3)\frac{3}{400}$$

$$\cdot\left[(3\cos^2\beta - 1 + \eta\sin^2\beta\cos 2\alpha)^2\frac{730}{729}\right.$$

$$\left.+(4+3\cos^2\beta+2\eta^2-\eta^2\cos^2\beta-2\eta\sin^2\beta\cos 2\alpha)\frac{20}{729}\right]. \quad (6.18)$$

This shows that the main contribution to M_2^{MAS} arises from the isotropic part, except for the case that the angle between both nuclei and the z-axis of the field gradient tensor is the magic angle. From Eq. (6.18), the narrowing factor can be obtained, where F is a function of β, α, and S:

$$\sqrt{\frac{M_2^{\mathrm{VanVleck}}}{M_2^{\mathrm{MAS}}}} = F\frac{\nu_S}{\nu_Q}. \quad (6.19)$$

The analytically obtained values agree with numerically calculated values of Böhm et al. [215] for $S = 3/2$ and $S = 5/2$ in the case of $\nu_Q < \nu_S$. For $\nu_Q \geqq \nu_S$ the perturbative approach cannot be applied. Böhm et al. [215] showed that for $\nu_Q > \nu_S$ there is only a small difference between M_2^{MAS} and M_2^{VanVleck}. For integer spins, ν_Q in Eq. (6.15) is not a common definition. But if we use this definition, the corresponding factors for integer spins can be calculated with Eq. (6.18). For coincident dipolar and quadrupolar principle axis systems the factor F can be approximated by

$$F \approx \frac{5}{2}\frac{1}{S} \quad (6.20)$$

with an error of less than 15%. Inserting F from Eq. (6.20) into Eq. (6.19) yields the narrowing factor

$$\sqrt{\frac{M_2^{\mathrm{VanVleck}}}{M_2^{\mathrm{MAS}}}} \approx \frac{5\nu_S}{2S\nu_Q} \quad (6.21)$$

Equation (6.21) shows that quadrupole effects limit the signal narrowing by MAS of spin-1/2 nuclei only for a high quadrupole frequency of the nucleus in the neighborhood of the spin-1/2 nuclei, $5\nu_s/2S\nu_Q < 100$. In this case, the NMR signal of the central transition of the quadrupole nucleus is strongly broadened, and an experimentally obtained MAS narrowing factor of the spin-1/2 nucleus in the neighborhood can be used for an estimate of the quadrupole frequency. An upper limit of C_{qcc} for ^{27}Al in the dehydrated zeolite H–Y was determined by this procedure [223].

The spin-1/2 spectra, calculated by using first-order perturbation theory, show for the individual values of m the typical quadrupole second-order lineshape, cf. Ref. [212]. Hence, the transmitted influence can be described as second-order quadrupole effect on NMR spectra of spin-1/2 nuclei [214].

7 Spin–Lattice Relaxation

The nuclear quadrupole moment couples to an electric field gradient (EFG) which, when time-dependent, causes transitions in the spin-system. Since the quadrupole interaction is typically much stronger than dipolar interactions, it is often more effective in relaxing the nuclear spins. Even for nuclear spins in a cubic environment, the time-dependent electric field gradient at the nuclear site is generally not zero. In solids, lattice vibrations cause a fluctuating EFG. The description of the lattice vibrations in terms of harmonic (or anharmonic) oscillators explains well the specific heat properties for crystals, as well as for amorphous substances at higher temperatures ($T \geqq 100$ K). However, for amorphous solids at lower temperatures, low-energy disorder modes (two-level systems, TLS) are assumed to explain the different specific heat capacity. Also, the spin–lattice relaxation of quadrupole nuclei has been found to be very different for amorphous solids at low temperatures compared with crystalline substances. Other processes responsible for a time-dependent EGF are the hopping of ions/defects which move electrical charge, or the motion of adsorbed species by moving electric dipolar moments.

Magnetic relaxation (fluctuating magnetic fields couple to the nuclear magnetic dipole moment) will not be considered in this review. For the coupling of the nuclear dipole moment to electron spins of paramagnetic impurities we refer to Khutsishvili [224]. Since semi-metals, metals or magnetic materials are also beyond the scope of this review, relaxation due to conduction electrons [225, 226] will not be considered, e.g. see Slichter [174] and Abragam [1]. The results of these theories can also be used to give an estimate for the relaxation of quadrupole nuclei, however, since the spin-temperature concept fails for a stronger quadrupole interaction, a detailed description of the relaxation process requires the solution of the master equations of the population numbers.

After a short review of the literature on spin–lattice relaxation of quadrupole nuclei in insulating solids, the basic theoretical aspects, necessary for the following more detailed discussion of the work published in this field, will be presented.

Pound [227] showed that the ^{23}Na nucleus in a single crystal of sodium nitrate is relaxed by a quadrupole interaction. Later, Andrew and Swanson [228] compared magnetic and quadrupolar relaxation mechanisms under the assumption of the spin-temperature concept. In a later paper, Andrew and Tunstall [49] extended the discussion to include suppressed spin-flipping due to quadrupole interaction. They solved the master equation, cf. Eq. (7.11) below, for two different initial conditions and calculated the eigenvalues for $I = 3/2$ and $5/2$ as a function of the ratio W_2/W_1, cf. definition of W_1, W_2 in Sect. 7.1. However, they did not discuss the nature of W_1 and W_2 in more detail. First theories for the calculation of W_1 and W_2 for the spin-phonon coupling in crystals, considering different processes in the lattice, were formulated by Van Kranendonk [229], Yosida and Moriya [230], Kondo and Yamashita [231], Kochelayev [232],

Mieher [233] and Wikner et al. [234]. They estimated the ratio W_2/W_1 to be not very different from unity. Tewari and Verma [235] extended the investigation of the master equation to $I = 7/2$ and 9/2 without focusing on the nature of W_i. Hughes [236] presented a detailed study of the master equation for $I = 3/2$ and different ratios of W_2/W_1, calculating the relaxation behavior for different initial conditions and single crystal orientations. Van Kranendonk and Walker [237] introduced the effect of anharmonic Raman processes as a main relaxation mechanism in crystals and discussed other models (using the spin-temperature approach). They showed that the Khutsishvili-process [1] does not lead to any relaxation if treated fully. Hughes and Reed [238] investigated the orientation dependence of the ratio W_2/W_1 for ^{23}Na ind sodium nitrate at room temperature and compared the results with theoretical predictions. They found that W_2/W_1 changes from about 0.85 to 1.25. Hikita and Tatsuzaki [239] investigated the temperature dependence of the relaxation of ^{23}Na in a single crystal of $AgNa(NO_2)_2$. They found experimental evidence that the observed T_1 cannot be explained by the spin-phonon coupling, however, by assuming that the fluctuating EFG is caused by the flipping of NO_2^- ions. Gordon and Hoch [240] discussed again, in more detail, the magnetization recovery in dependence on the ratio W_2/W_1 for $I = 3/2$ to $I = 9/2$. In particular, they discussed the initial conditions for pulsed NMR experiments.

After the discovery that at low temperatures, the specific heat in noncrystalline solids is much greater than in crystalline solids, the relaxation of quadrupole nuclei in such systems became the focus of much research. Rubinstein and Taylor [241] found by comparing crystalline and amorphous As_2S_3 that the relaxation of ^{75}As in the amorphous sample was indeed much faster. Szeftel and Alloul [242] measured the T_1 of ^{11}B and ^{23}Na in amorphous and crystalline samples of B_2O_3, $(Na_2O)_{0.3}(SiO_2)_{0.7}$ and $Na_2B_4O_7$ in the temperature range of 1.2 K to 300 K, and found that the relaxation rate in the amorphous samples was up to two orders of magnitude faster than in crystalline samples, and was nearly frequency independent. Rubinstein et al. [243], who investigated the temperature and frequency dependence of the ^{11}B relaxation in amorphous and crystalline B_2O_3 at temperatures between 150 K and 500 K, found a minimum for the relaxation at about 300 K, and were able to present a theoretical explanation. Reinecke and Ngai [244] discussed the different processes of relaxation in amorphous solids (glasses) up to 100 K, including all experimental data published up to 1975. They found that at low temperatures the Raman process involving two disorder modes is the most effective, in contrast to the explanation of Szeftel and Alloul [242] who claimed that the Raman processes involving a disorder mode and a phonon is responsible for the relaxation. Later, Szeftel and Alloul [245] presented a detailed theoretical study of quadrupole relaxation in amorphous materials, and, together with some new experimental results, they reviewed all experimental data available up to 1978. All the observed relaxation rates (for ^{11}B and ^{23}Na) have a $T^{1+\eta}$ dependence ($\eta < 1$). Slight differences can be explained by a fluctuating EFG due to moving Na-ions. The relaxation rate is only slightly field dependent. The authors analyzed the different processes, and found that neither the Raman process involving two disorder modes (for which the relaxation rate vanishes, opposed to previous work [244]), nor the Raman process involving one phonon and one disorder mode can explain the experimental data. However, the formerly discarded process involving only a single disorder mode can explain the experimental data, provided that the disorder mode is coupled to other disorder modes, rather than to the phonons. A point-charge model for two positions of a bridging oxygen atom, which represents the disorder mode (two-level system, TLS), yielded satisfactory agreement for the ^{11}B spin–lattice relaxation time in B_2O_3.

Avogadro et al. [246] investigated the ^{11}B spin–lattice relaxation in ionic glassy conductors. At temperatures above 100 K, the data can be explained by the (thermally activated) reorientational motion of BO_4 groups. At lower temperatures, the almost linear relation between $1/T_1$ and T is readily explained by disorder modes. However, in order to account for the observed, complicated frequency dependence of the relaxation, they developed a model of two different kinds of disorder modes (one in good contact with the phonon bath, the other isolated).

Bjorkstam and Villa [247, 248] developed a theory for quadrupole relaxation due to two-dimensional motion of ions and found a good overall agreement for the ^{27}Al and ^{23}Na relaxation in Na β-alumina single crystals. Greenbaum et al. [249, 250] investigated ^{27}Al and ^{23}Na relaxation in Na β-alumina at even lower temperatures. Below 55 K they found that the relaxation can be explained by energetically low-lying excitations (corresponding to two-level systems). Becker [251] discussed the relaxation induced by the motion of lattice defects in solids. Tokuhiro et al. [252] measured the ^{7}Li relaxation time in several hydrated zeolites, where the Li^+ cation relaxation is described by the viscous solution approach (quadrupole relaxation in liquids, cf. [1], p. 313 ff.). Haase et al. [253, 254] addressed the ^{27}Al relaxation in zeolites (100 K–400 K) and showed that for hydrated zeolites the typically short T_1 is caused by the fluctuations of the EFG due to the electric dipole moment of adsorbed water molecules, and for dehydrated zeolites, the observed relaxation behavior seems to be in accordance with the suggested models of the motion of cations. Towta and Hughes [255] studied the ^{23}Na relaxation for a single crystal of $NaNO_2$ at various temperatures. Below 145 K, they find the relaxation to be magnetic, but at higher temperatures (up to 450 K) quadrupole relaxation is dominant. A theoretical analysis allows the determination of the different transition rates, $W_{1,2}$, which are related to the crystal structure. Haase et al. [256] investigated the temperature dependent spin–lattice relaxation times of ^{17}O, ^{23}Na, ^{27}Al, and ^{71}Ga in a series of hydrated zeolites. They found that there is no correlation between T_1 and the static part of the electric field gradient. However, for a given structure and amount of hydration, the observed values of T_1 of the different nuclei is readily explained by assuming the fluctuating EFG at each nuclear site is due to the electric dipolar moment of the absorbed water plus the Sternheimer antishielding effect for each atom.

7.1 Basic Considerations

Spin–lattice relaxation, described by a characteristic time constant T_1, is an irreversible energy transfer from the nuclear spin-system to the "lattice" (an energy reservoir), in which the non-equilibrium magnetization approaches its thermodynamic equilibrium value. The "lattice" can be any system with a much larger heat capacity than that of the nuclear spin-system. The large heat capacity of the "lattice" ensures that the lattice temperature can be assumed to be constant during the relaxation process. Furthermore, it is assumed that dissipative processes within the lattice destroy any coherences caused by the contact with the spin-system in a characteristic time τ, the correlation time of the lattice. In insulating solids, the physical nature of the "lattice" can be described by ordinary lattice vibrations (phonons), or by two-level systems (disorder modes) for amorphous solids and temperatures below $T = 100\,\mathrm{K}$. Adsorbed atoms or molecules in crystals or porous materials can also represent the "lattice" if they have a large heat capacity (e.g. for hydrated zeolites most of the heat capacity stems from adsorbed water molecules).

For spin–lattice relaxation of quadrupole nuclei in solids the static part of the quadrupole interaction, which causes a shift of the Zeeman levels, is most often large enough to inhibit the spin-exchange between different transitions. Thus, a spin-temperature will not be established as fast as for the case of a vanishing quadrupole coupling, and, the time-dependence of the magnetization built-up is different for each transition. This is inconsistent with the BPP-theory [174] of relaxation, in which the spin-temperature concept is invoked. If the static quadrupole interaction prohibits spin-flipping, the degree of freedom for the population numbers of the $2I + 1$ levels is $2I$ since the total number of spins is a constant. Assuming exponential decays, $2I$ eigenvalues (T_1) are to be expected for the decay of the magnetization of a single transition. For crude estimates of the spin–lattice relaxation the effect of the static quadrupole interaction, which suppresses spin-flipping, may be discarded. However, for more detailed studies the above mentioned effects have to be considered.

The total Hamiltonian for the description of the spin–lattice relaxation is

$$\mathscr{H} = \mathscr{H}_L + \mathscr{H}_Q(t) + \mathscr{H}_R, \tag{7.01}$$

where $\mathscr{H}_L$, $\mathscr{H}_R$ represent the time-independent Hamiltonians for the spin-system (Zeeman or Larmor interaction) and the lattice (reservoir), respectively. $\mathscr{H}_Q(t)$ denotes the time-dependent coupling between the spin-system and the energy reservoir. For the eigenfunctions of the combined system, $\mathscr{H}$, we use the product functions $|m, r\rangle = |m\rangle|r\rangle$, where

$$\mathscr{H}_L|m\rangle = E_m|m\rangle \quad \text{and} \quad \mathscr{H}_R|r\rangle = E_r|r\rangle. \tag{7.02}$$

The interaction representation is chosen by separating the time-independent part of $\mathscr{H}$, $\mathscr{H}_0 = \mathscr{H}_L + \mathscr{H}_R$. Then, we have in the interaction representation: $\mathscr{H}_{R,i} =$

$\mathscr{H}_R$, $\mathscr{H}_{L,i} = \mathscr{H}_L$, and

$$\mathscr{H}_{Q,i}(t) = \exp\left\{-\frac{i}{\hbar}(\mathscr{H}_L + \mathscr{H}_R)t\right\}\mathscr{H}_Q(t)\exp\left\{\frac{i}{\hbar}(\mathscr{H}_L + \mathscr{H}_R)t\right\}. \qquad (7.03)$$

For small changes in the population of the levels $|m,r\rangle$, $|m',r'\rangle$ one may write for the transition probability, P, for the transition $|m,r\rangle \to |m',r'\rangle$, using Dirac's first-order, time-dependent perturbation theory

$$\langle m',r'|P|m,r\rangle = \frac{1}{\hbar^2}\left|\int_0^t \langle m'r'|\mathscr{H}_{Q,i}(t')|mr\rangle \mathrm{d}t'\right|^2. \qquad (7.04)$$

The transition rate, $\langle m',r'|W|m,r\rangle \equiv W_{mr,m'r'} = \frac{\partial\langle m'r'|P|mr\rangle}{\partial t}$, is the rate by which the photon necessary to accomplish the spin-flip in the nuclear spin-system is absorbed (emitted) by the lattice. If $\mathscr{H}_Q$ is time independent, and the lattice states are close enough, the well-known formula results

$$\langle m'|W|m\rangle = \frac{2\pi}{\hbar}\left|\langle m'|\mathscr{H}_Q|m\rangle\right|^2 \rho(E), \qquad (7.05)$$

where $\rho(E)$ represents the energy density of the considered processes in the lattice. More generally, it follows from Eq. (7.04)

$$W_{mr,m'r'} = \frac{1}{\hbar^2}\int_0^t \{\langle m'r'|\mathscr{H}_{Q,i}(t)|mr\rangle\langle m'r'|\mathscr{H}_{Q,i}(t')|mr\rangle$$
$$+ \langle mr|\mathscr{H}_{Q,i}(t)|m'r'\rangle\langle m'r'|\mathscr{H}_{Q,i}(t')|mr\rangle\}\,\mathrm{d}t'. \qquad (7.06)$$

It is usually assumed that the ensemble average, $\overline{\langle m'r'|\mathscr{H}_{Q,i}(t)|mr\rangle\langle m'r'|\mathscr{H}_{Q,i}(t')|mr\rangle}$, vanishes for $|t-t'| \to \infty$. This means, the involved transitions become phase-incoherent after a time difference, $\tau = t - t'$. In addition, the correlations are assumed to be stationary, i.e., in the ensemble average the products of the matrix elements in Eq. (7.06) are a function of the time difference, $t - t'$, only. For $\mathscr{H}_Q(t)$ we consider the time-dependent quadrupole Hamiltonian, and that in the ensemble average the matrix elements of $\mathscr{H}_{Q,i}$ vanish, $\overline{\langle m'r'|\mathscr{H}_{Q,i}(t)|mr\rangle} = 0$, or, the mean value is considered to be part of the static quadrupole interaction. $\mathscr{H}_{Q,i}$ follows from Eq. (1.56), as

$$\mathscr{H}_{Q,i}(t) = C_Q \sum_{q=-2}^{+2} (-1)^q T_q V_{-q,i}(t)\exp\{iq\omega t\}, \qquad (7.07)$$

where $V_{-q,i}(t) = \exp\left\{-\frac{i}{\hbar}\mathscr{H}_R t\right\}V_{-q}(t)\exp\left\{\frac{i}{\hbar}\mathscr{H}_R t\right\}$ are the operators for the components of the electric field gradient. As is seen with Eqs. (7.06) and (7.07), the spin-dependent part of the quadrupole Hamiltonian restricts the transitions in the spin-system according to the matrix elements

$$\langle m'|T_0|m\rangle = \frac{3m^2 - I(I+1)}{\sqrt{6}}\delta_{m',m}$$

$$\begin{aligned}\langle m'|T_{\pm 1}|m\rangle &= -(2m+1)W_m\delta_{m'm+1} + (2m-1)W_{m-1}\delta_{m'm-1}\\ &\equiv \mp(2m\pm 1)W_m^{\pm}\delta_{m'm\pm 1}\\ \langle m'|T_{\pm 2}|m\rangle &= 2W_{m+1}W_m\delta_{m'm+2} + 2W_{m-2}W_{m-1}\delta_{m'm-2}\\ &\equiv 2W_{m\pm 1}^{\pm}W_m^{\pm}\delta_{m'm\pm 2},\end{aligned} \tag{7.08}$$

where

$$W_m = \tfrac{1}{2}\sqrt{I(I+1) - m(m+1)} \equiv W_m^+,\ W_m^- \equiv W_{m-1} = \tfrac{1}{2}\sqrt{I(I+1) - m(m-1)},$$

$$\text{or } W_m^{\pm} = \tfrac{1}{2}\sqrt{(I\mp m)(I\pm m+1)}.$$

Since we are calculating the transition rates, $m' \neq m$, one is left with single and double quantum transitions. The matrix elements $\langle m'|T_{\pm q}|m\rangle$, Eq. (7.08), are introduced into Eq. (7.06), and with the above mentioned assumption one can write

$$\begin{aligned}W_{mr,m'r'} = \frac{2}{\hbar^2}C_Q^2\{&(2m\pm 1)^2(W_m^{\pm})^2\tilde{W}_1(r,r')\delta_{m'm\pm 1}\\ &+ (2W_m^{\pm})^2(W_{m\pm 1}^{\pm})^2\tilde{W}_2(r,r')\delta_{m'm\pm 2}\},\end{aligned} \tag{7.09}$$

with

$$\begin{aligned}\tilde{W}_n(r,r') = \int_{-\infty}^{+\infty} &\exp\{i[\pm n\omega_L + (E_r - E_{r'})/\hbar]\tau\}\\ &\cdot Re\,\overline{\{\langle r'|V_{\pm 2}(t)|r\rangle\langle r|V_{\mp 2}(t-\tau)|r'\rangle\}}\,d\tau.\end{aligned}$$

Inserting $C_Q = \dfrac{eQ}{2I(2I-1)}$ and $W_m^{\pm}$ from Eq. (7.08), and with $W_n(r,r') = \dfrac{e^2Q^2}{4I\hbar^2}\tilde{W}_n(r,r')$ it follows

$$\begin{aligned}W_{mr,m'r'} = &\frac{(2m\pm 1)^2(I\mp m)(I\pm m+1)}{2I(2I-1)^2}W_1(r,r')\delta_{m'm\pm 1}\\ &+ \frac{(I\mp m)(I\pm m+1)(I\mp m-1)(I\pm m+2)}{2I(2I-1)^2}W_2(r,r')\delta_{m'm\pm 2}.\end{aligned} \tag{7.10}$$

The definition of W_1 and W_2, as given by Eq. (7.10), differs in the literature. The one given in Eq. (7.10) corresponds to those used by Yosida and Moriya [230] Andrew and Tunstall [49], Hughes and Reed [238]. It is also common [253, 256] to extract the spin-quantum number I from the definition of $W_{1,2}$ resulting in a change of I to I^2 in the numerators of Eq. (7.10). The integral in Eq. (7.09) represents a spectral density, the real part, $Re\{\ \}$, is the correlation function of the processes in the lattice. In order to find the total transition rate corresponding to a single or double quantum process in the spin-system, $W_{1,2}(r,r')$ has to be multiplied by the density of the states and summed over r, r'.

So far, we have only been concerned with the expressions for transitions in the spin system and the lattice. In order to describe the behavior of the spin system (the time-dependence of the magnetization) one can start with the master equation of the population numbers, N_m, the average number of spins populating the Zeeman level, m,

$$\frac{dN_m}{dt} = \sum_{n=-I}^{I} W_{nm}(N_n - N_{n0}) - (N_m - N_{m0}) \sum_{n=-I}^{I} W_{mn} \tag{7.11}$$

where N_{m0} is the thermal equilibrium population of level E_m, and, W_{nm} denotes the transition rate $n \rightarrow m$ in the spin-system for all possible transitions in the lattice, and follows from Eq. (7.09). Equation (7.11) is a system of coupled linear differential equations. The general solution is of the form $N_m(t) - N_{m0} = \sum_{k=-I}^{I-1} C_k(t=0) \exp\{-\lambda_k t\}$. By introducing the differences in the population numbers, $n_m = N_m - N_{m+1} (n_{m0} = N_{m0} - N_{m0+1})$, one can write

$$n_m(t) - n_{m0} = \sum_{k=-I}^{I-1} c_k(t=0) \exp\{-\lambda_k t\}. \tag{7.12}$$

Since the total number of spins is a constant, the system has only 2I non-vanishing eigenvalues, λ_k. The initial conditions are given by the excitation, e.g., a $\pi/2$-pulse at the carrier frequency, ω_L (Larmor frequency), will yield: $N_n(t=0) = -N_{-n}(t=0)$. As a result of special initial conditions, some of the constants c_k will vanish. The solution of the eigenvalue problem for $I > 3/2$ is not simple since higher order equations have to be solved for every ratio of W_2/W_1. In addition, the constants c_k have to be determined using the initial conditions and the eigenvectors for every W_2/W_1. These procedures can easily be accomplished using computers, making any further discussion unnecessary. For the case of a classical lattice and nonselective excitation of the central transition we will give approximations for the T_1 minimum. References to these problems will be given further below.

The excitation of the spin-system affects the relaxation behavior: The initial conditions influence the constants, c_k, in Eq. (7.12). Thus, only the eigenvalues, λ_k, for which the c_k do not vanish are important for the relaxation process. This may cause a change in the multi-exponential character for the measured T_1. Especially, since the quadrupole interaction is temperature dependent, the excitation/measurement of T_1 is not trivial.

The excitation has been discussed in Sect. 2. Here we would only like to mention that for powders the saturation (or inversion) of the central transition can always be achieved by irradiation of a selective $\pi/2$ (or π) pulse. The saturation of the whole spin-system with a large quadrupole interaction cannot be easily achieved by the application of a train of pulses, cf. Fig. 2.1; the total length of the pulse train has to be long compared with T_1. The case of non-selective excitation applies also, if a sample is in thermodynamic equilibrium outside of the static magnetic field and is placed into a magnetic field within a time $t \ll T_1$.

7.2 *Spin-Phonon Coupling in Crystals*

Relaxation due to a fluctuating EFG caused by lattice phonons involves processes by which the photons, necessary for nuclear spin-flips, are emitted/absorbed by the lattice. The effectiveness of such processes depends on the spin–lattice coupling and the lattice properties, and both have to be expressed in quantum mechanical terms. Van Kranendonk [229] presented the first theoretical approach for these processes. In a later paper, Van Kranendonk and Walker [237] introduced a new mechanism, and reviewed the hitherto literature, see also Abragam [1]. The electric field gradients, V_q, in Eq. (7.09) are usually expanded in terms of the components of the mechanical strain tensor, ε_k. Symbolically one may write, $V_q \equiv V = V_0 + V_k\varepsilon_k + \varepsilon_k V_{kl}\varepsilon_l + \ldots$ The strains, ε_k, are derivatives of the displacements of the nuclei with respect to the coordinates, which can be calculated from the quantized displacement operators given by the lattice Hamiltonian. The components of the EFG in the basis of the strain tensor components depend on the actual lattice structure. However, it is usually assumed that these components are comparable to the magnitude of the static electric field gradient, a rather crude estimate. The lattice Hamiltonian, which depends on the actual structure as well, is approximated by its harmonic and anharmonic part. The static part of the quadrupole interaction is neglected. Nevertheless, with the mentioned assumptions and Debye's model of the phonon density, it is possible to find the field and temperature dependence for the several terms, and estimate their magnitudes using Eq. (7.05). Since the energies for the two possible nuclear spin-flips, $\hbar\omega_L$ or $2\hbar\omega_L$, are very small quantities compared with the lattice, to a good approximation, the density of phonons can be regarded as being constant over an energy range of $2\hbar\omega_L$, hence, $W_1 \approx W_2$. The first term of the expanded EFG is the constant (static) term. The second term is proportional to ε_k, and thus involves only one phonon. In this *direct process*, the lattice absorbs (emits) the photon necessary for the nuclear spin-flip. One finds for the relaxation rate [229], $W_{m,m\pm1} \approx W_{m,m\pm2} \approx \dfrac{1}{T_1}$,

$$\frac{1}{T_1} \approx C_1\omega_L^2 T \quad \text{direct process,} \tag{7.13}$$

where C_1 is a constant and T is the absolute temperature. Although, the probability for multiphonon processes in the lattice decreases rapidly with the number of involved phonons, it has been shown [229] that the direct process can be neglected compared with the two-phonon process (*Raman process*), third term in the expansion for V_q, where the photon, necessary for the nuclear spin-flip, is absorbed (emitted) by a two-phonon process in the lattice. The conservation of the energy required $n\hbar\omega_L = \hbar\omega_1 + \hbar\omega_2$. The relaxation rate becomes

$$\frac{1}{T_1} \approx C_2 \frac{T^2}{\Theta} \quad (T \gtrsim \Theta) \quad \text{Raman process,} \tag{7.14}$$

where Θ is the Debye temperature of the lattice, and C_2 is a constant. At lower temperatures the dependence is more complicated, and finally becomes

$$\frac{1}{T_1} \approx C_2' \frac{T^7}{\Theta^6} \quad (T \ll \Theta) \quad \text{Raman process,} \tag{7.15}$$

where C_2' is a constant.

Lattice vibrations in a crystal are always more or less anharmonic (a necessity for the heat conductivity). Van Kranendonk and Walker [237] showed that for an anharmonic crystal another process, the *anharmonic Raman process*, can be very important. The anharmonic Raman process involves a direct process, where the created (or absorbed) phonon is the product of an inelastic scattering of two phonons. This process shows the same temperature dependence as given by Eqs. (7.14) and (7.15).

It should be mentioned that the spin–lattice relaxation times, T_1, as given by Eqs. (7.13–7.15) are crude estimates, the temperature and field dependence are the main features of the above equations. Relaxation due to a spin-phonon coupling has often been found to be too weak to account for experimentally measured relaxation rates, especially at moderate and higher temperatures. This is probably due to the lack of knowledge of the actual fluctuations of the EFG in the crystal. Also, other, more effective relaxation mechanisms like moving ions/defects, reorientation of polar groups or adsorbed species, and relaxation due to paramagnetic impurities, often mask the relaxation induced by the spin-phonon coupling.

7.3 Spin Relaxation in Amorphous Materials

In addition to regular phonons, in amorphous materials disorder modes are present, which are assumed to be two-level systems, TLS. In this case, the electric field gradient, V_q, is expanded not only in terms of the strain tensor elements, ε_i, but also in terms of the disorder modes (since the latter is a two-level system, it can be described by a pseudospin σ[47]). The EFG is then given by the symbolic sum, $V_q \equiv V = V_0 + V_k\varepsilon_k + \varepsilon_k V_{kl}\varepsilon_l + V_i\sigma_i + \sigma_i V_{ij}\sigma_j + \varepsilon_k V_{ki}\sigma_i + \ldots$. The first, static term does not contribute to the relaxation. The second term is the *direct process*, and as already mentioned is usually not important for the relaxation. The third term leads to Raman processes involving two phonons, and does not fit the experimental results of an almost linear temperature dependence of the relaxation rate for $T < 100\,\text{K}$. Reinecke and Ngai [244] suggested a Raman process involving two disorder modes, $\sigma_i V_{ij} \sigma_j$. Starting from Eq. (7.05) and the spin-temperature approximation they estimate the relaxation rate, $\frac{1}{T_1} \sim W_2$,

$$\frac{1}{T_1} \approx \frac{2\pi}{\hbar^2} T_{\mathrm{m}}^2 \int_0^{E_m} |V|^2 \rho(E)^2 \left[1 + \cosh\frac{E}{k_{\mathrm{B}}T}\right]^{-1} \mathrm{d}E, \tag{7.16}$$

where T_{m} corresponds to one of the matrix elements of T_{m} in Eq. (7.08), V is the matrix element of the EFG, $\rho(E)$ is the density of disorder modes, E is the energy splitting of the disorder modes. The maximum energy splitting in the material is $E_{\mathrm{m}}(E_{\mathrm{m}} \gg k_{\mathrm{B}}T$ for $T \lesssim 100\,\mathrm{K})$. Then, by assuming an almost constant energy density of the disorder modes, $\rho(E) = \rho_0 E^{\eta}$, where $\eta < 1$ (as found experimentally at low temperatures), and that the derivative of the EFG with respect to the strains is independent on the energy, E, they find

$$\frac{1}{T_1} \approx T^{1+\eta} \quad \text{(Raman process involving two disorder modes).} \tag{7.17}$$

Szeftel and Alloul [242] had proposed the process, $\varepsilon_{\mathrm{k}} V_{\mathrm{ki}} \sigma_{\mathrm{i}}$, where a phonon and a disorder mode are involved, but as they pointed out 3 years later [245], this process leads to the wrong temperature dependence. In the latter paper they found that the Raman process, $\sigma_{\mathrm{i}} V_{\mathrm{ij}} \sigma_{\mathrm{j}}$, yields vanishing matrix elements, contradicting the suggestion of Reinecke and Ngai [244]. The process represented by the term, $V_{\mathrm{i}}\sigma_{\mathrm{i}}$, which involves only one TLS, and which was discarded by Szeftel and Alloul [242] in 1975, was reintroduced by the same authors in 1978 [245]. Assuming $W_1 \approx W_2$, it follows from Eq. (7.10), by summing over the lattice

$$\frac{1}{T_1} \approx \frac{1}{10} \frac{e^2 Q^2}{\hbar^2} \int_{-\infty}^{+\infty} V_{\mathrm{i}}^2 \langle \sigma_{\mathrm{i}}(t)\sigma_{\mathrm{i}}(0) \rangle \exp\{-\mathrm{i}\omega_{\mathrm{L}} t\}\, \mathrm{d}t, \tag{7.18}$$

where $\langle \sigma_{\mathrm{i}}(t)\sigma_{\mathrm{i}}(0) \rangle \equiv \mathrm{tr}_{\mathrm{r}} \left\{ \exp\left[\frac{\mathrm{i}}{\hbar} \mathscr{H}_{\mathrm{R}} t\right] \sigma_{\mathrm{i}}(t) \exp\left[-\frac{\mathrm{i}}{\hbar} \mathscr{H}_{\mathrm{R}} t\right] \sigma_{\mathrm{i}} \rho_{\mathrm{R}}(0) \right\}$ includes the sum over all lattice states $|r\rangle$ with the equilibrium distribution $\rho_{\mathrm{R}}(0) = \exp\left\{-\frac{\mathscr{H}_{\mathrm{R}}}{k_{\mathrm{B}}}\right\} \Big/ \mathrm{tr}_{\mathrm{r}} \left\{ \exp\left[-\frac{\mathscr{H}_{\mathrm{R}}}{k_{\mathrm{B}}}\right] \right\}$. Szeftel and Alloul [245] evaluated the correlation function $\langle \sigma_{\mathrm{i}}(t)\sigma_{\mathrm{i}}(0) \rangle$ under the assumption that $\mathscr{H}_{\mathrm{R}}$ includes the harmonic lattice, and a coupling between the two-level systems. They concluded that the relaxation rate should have the form

$$\frac{1}{T_1} \approx \nu_{\mathrm{Q}} \tau_{\mathrm{z}} \left[\cosh \frac{E}{2k_{\mathrm{B}}T}\right]^{-2}, \tag{7.19}$$

where τ_{z} is the correlation time of the disorder modes, E its energy. After averaging over different TLS energies, E, they find basically the same temperature dependence as given by Eq. (7.17). Both approaches seem to describe the spin–lattice relaxation. The temperature dependence, the independence on the Larmor frequency and the magnitude of the relaxation rate is explained.

7.4 *Activated Processes*

Moving electric charges and dipolar moments cause fluctuations of the electric field gradient. In this case E_r and $E_{r'}$ are assumed to be infinitely close and the V_q in Eq. (7.09) can be considered as classical components of the *EFG* tensor. We assume the following correlation function

$$\overline{V_{\pm q}(t)V_{\mp q}(t-\tau)} = V_0^2 \exp\{-|\tau|/\tau_c\}, \tag{7.20}$$

where V_0 is the maximum EFG at the nuclear site (which might include antishielding effects) and τ_c is the correlation time of the motion. Then, the transition rates W_1 and W_2 from Eq. (7.10) can be calculated as

$$W_{m,m'} = \frac{(2m \pm 1)^2(I \mp m)(I \pm m + 1)}{2I(2I-1)^2} \frac{e^2Q^2V_0^2}{2I\hbar^2} \frac{\tau_c}{1+(\omega_L\tau_c)^2} \delta_{mm\pm 1}. \tag{7.21}$$

An often used assumption about the temperature dependence of the correlation function is the Arrhenius equation

$$\tau_c = \tau_0 \exp\{E_A/RT\}, \tag{7.22}$$

where E_A is the activation energy, and $R = N_0 k_B$ the gas constant and T the absolute temperature. With the Eqs. (7.20) and (7.21) and the knowledge of the initial conditions, the master equation, Eq. (7.12), can be solved, and the magnetization recovery, T_1, E_A and τ_0 can be determined.

It has been shown [256] that for powders the selective excitation and detection of the central transition gives a nearly single-exponential relaxation, hence, the minimum T_1 can be estimated by choosing the dominating eigenvalue in the master equation. For selective saturation and detection of the central transition one finds [256]

$$T_{1\min} = k_1(I)\frac{\hbar\omega_L}{e^2Q^2V_0^2} = k_2(I)\frac{\nu_L}{\nu_Q^2} \tag{7.23}$$

where $k_{1(2)}(I) = 3/2\,(6), 10/3\,(4000/27), 21/4\,(1029)$, and $36/5\,(20736/5)$ for $I = 3/2$, 5/2, 7/2, and 9/2, respectively.

8 Intensity Measurements

The observed NMR intensity is a function of the spectrometer, as well as of the properties of the spin-system, the temperature and external fields. Thermodynamic and macroscopic considerations have been neglected, so far, since it is convenient for the theoretical treatment, and, the free induction decay (FID) was defined in Eq. (1.51) to be 1 after a nonselective $\pi/2$ pulse. However, in order to compare the number of different spins (in different samples), macroscopic effects have to be considered.

It is well known [1], that the equilibrium magnetization due to N nuclear spins, I, in a static magnetic field, B_0, which experience only a Zeeman interaction, $\mathscr{H}_L = -\sum_{k=1}^{N} \gamma\hbar I_z^{(k)}$, is given by

$$M_z = \frac{\operatorname{tr}\left\{\exp\left[-\frac{\mathscr{H}_L}{k_B T}\right]\gamma\hbar I_z\right\}}{\operatorname{tr}\left\{\exp\left[-\frac{\mathscr{H}_L}{k_B T}\right]\right\}} = \frac{N\gamma^2\hbar^2 I(I+1)}{3kT} B_0, \tag{8.01}$$

where the high temperature approximation, $\frac{\gamma\hbar B_0}{k_B T} \ll 1$, has been used. An applied rf pulse will influence this magnetization, and an oscillating magnetic field, M_s, appears along the axis of the solenoid. Since this magnetic field, M_s, oscillates with the Larmor frequency, $M_s = M_{s0}\exp(i\omega_L t)$, the voltage induced in the coil, which is proportional to $\partial M_s/\partial t$, has the amplitude, $\omega_L M_{s0} = \gamma B_0 M_{s0}$. For a given excitation, M_{s0} is always proportional to M_z, therefore, the FID given by Eq. (1.51) should be redefined, as

$$G(t) = G_0 \frac{\operatorname{tr}\{\rho_i(t) I_+\}}{\operatorname{tr}\{I_x^2\}} = G_0 \frac{\langle I_- | \rho_i(t)\rangle}{\langle I_x | I_x\rangle}, \tag{8.02}$$

where

$$G_0 = \frac{N\hbar^2 I(I+1)}{3kT} \gamma^3 B_0^2.$$

Equation (8.02) shows the dependence of the NMR intensity in a more general way, however, it does not account for properties of the probe, such as the quality factor, filling factor of the coil or frequency dependence of amplification and noise. That is why, it is not possible to compare NMR intensities of different nuclear spins under very different conditions with high accuracy, however, in some cases, e.g., comparison of ^{23}Na and ^{27}Al intensity in a given sample, it might be useful.

Since for pulse spectrometers the probe is usually tuned and matched upon changing frequency or sample, it should be mentioned that in this case, the amplitude of the rotating magnetic rf field, ω_{rf}, is porportional to the squared root of the power, P, and the quality factor, Q, of the circuit, $\omega_{rf} \propto \sqrt{PQ} \propto U\sqrt{Q}$, where U is the voltage of the transmitter. Therefore, both the rf voltage transmitted into the coil and the NMR induction in the coil have to be multiplied by $\sqrt{Q}$. Since the quality factor of the probe may change upon changing the sample, probes with a low quality factor should be used for intensity measurements.

A crucial point for the comparison of intensities is the rf inhomogeneity. Since the rf field varies strongly over the volume of the coil, the rf amplitude, ω_{rf}, as well as the voltage induced by the magnetization will be different for the nuclei throughout the sample unless the sample volume is small compared with the size of the coil. Therefore, when comparing intensities, both samples should be small or large compared with the size of the coil, or, they should have exactly the same size and position with respect to the coil.

The intensities to be compared should be similar in order to avoid errors caused by a nonlinearity of the amplification in the detection channel. Also, since the errors connected with the detection of a NMR signal are expected to be similar for similar lineshapes, the signals to be compared should have a similar linewidth. The FID for a Gaussian line is

$$G(t) = G(0) \exp\left\{-\frac{\pi^2}{4 \ln 2}(\Delta t)^2\right\}, \tag{8.03}$$

where Δ is the full width at half height (FWHH) in frequency units. By expanding Eq. (8.03), it is seen that for a Gaussian line a small loss of intensity during the dead time of the spectrometer can be approximated by $\frac{\pi^2}{4 \ln 2}(\Delta t)^2 \approx 3.56(\Delta t)^2$. For a linewidth of $\Delta = 10\,\text{kHz}$ and a dead time $t = 10\,\mu\text{s}$ the loss of intensity is about 3.56%.

In order to avoid saturation effects, the repetition time for data accumulation has to be bigger than about $5\,T_1$.

For powder spectra of quadrupole nuclei, the excitation, as discussed in Sect. 2, can be limited by both the duration of the exiting pulse (pulse sequence) and/or the bandwidth of the probe. With commercial solid-state NMR spectrometers a uniform excitation of about 1 MHz (single pulse experiment) can be achieved. This corresponds to a quadrupole frequency $\nu_Q \approx \frac{1000}{(2I-1)}$ kHz. If the quadrupole frequency of the nuclei is *small* compared with this value, excitation effects can be neglected, however, the detection may still be a problem and will be discussed in Sect. 8.1. In Sect. 8.2 the case $\|\mathcal{H}_{rf}\| \leqq \|\mathcal{H}_Q\|$ will be considered more closely. Here, the influence of the quadrupole interaction cannot be neglected during the application of the pulse, and the determination of intensities becomes rather complicated. Section 8.3 describes problems of intensity measurement which occur for any excitation if the MAS technique is applied.

8.1 *Small Quadrupole Coupling*

In the case of nonselective excitation by a single hard pulse, $\|\mathcal{H}_Q\| \ll \|\mathcal{H}_{rf}\|$, one can simply compare the observed intensity with a standard sample, provided that the loss of signal during the dead time of the spectrometer can be neglected. However, if one is forced by the initial signal loss to employ other than single pulse experiments, e.g. spin-echo experiments, the quadrupole interaction can cause complications. A hard pulse means that the influence of the quadrupole coupling during pulsing can be neglected, but for multiple pulse experiments, this is not necessarily the case during the time between the pulses. If one can ensure that the loss of signal immediately before acquisition is negligible, the intensity measurement by a single pulse experiment is recommended, since it is less sensitive to a slight missetting of the duration of the pulse.

If the quadrupole interaction is of no importance at all, a chemical shift anisotropy and heteronuclear dipolar interaction are most likely to cause wide resonance lines, forcing the application of a spin-echo sequence. A $\pi/2-\pi$ spin-echo experiment can be used since it refocuses the dephasing due to those interactions. The effect of a combined homonuclear and heteronuclear coupling on spin-echo measurements has been described by Mansfield [74]. The strength of these interactions have to be known in order to estimate the loss of the spin-echo amplitude. However, homonuclear dipolar interactions usually account for a linewidth contribution of less than 10 kHz. Hence, intensity measurements are not much affected, and a single pulse experiment or a $\pi/2-\pi$ spin-echo (with a short pulse delay) can be employed for intensity measurements.

If the quadrupole interaction exceeds the other interactions (hard pulses still assured) a two-component FID can be observed. The first, more rapid decay of the FID (corresponding to the broad components in the Fourier transform spectrum) is caused by the dephasing of the satellite transitions, whereas the slower decay (narrow component in the Fourier transform spectrum) belongs to the central transition intensity. Now, a 90° phase shifted $\pi/2-\pi/2$ echo sequence *partly* refocuses the dephasing of the satellites, causing a somewhat slower decay of the first more rapid part in the spin-echo decay compared with the one-pulse FID. However, some satellite intensity may still be missing after the ringing. In addition, the effects of a chemical shift anisotropy and of heteronuclear dipolar couplings are not removed by using a 90° phase shifted $\pi/2-\pi/2$ pulse sequence. Therefore, in case of a suspected loss of intensity during the ring down time, the use of the a $\pi/2-\pi$ spin-echo sequence is recommended for the intensity measurements [257]: For the $\pi/2-\pi$ spin-echo sequence the effect of the quadrupole interaction is the same as for the FID, which means the dephasing due to quadrupole interaction is not refocussed. The fast decaying component has a time constant of about $1/\nu_Q$. However, the effects of the chemical shift anisotropy and heteronuclear dipolar coupling on the spin-echo amplitude are removed. Now, recording several spin-echo amplitudes for different pulse distances, the intensity of the central transition can be determined with high accuracy. The hereby determined intensity (nonselective excitation) has to be multiplied by the factor *f*,

$$f=\frac{8\cdot I(I+1)}{3\cdot(2I+1)}, \qquad (8.04)$$

which accounts for the difference of the central transition intensity compared with the total intensity for nonselective excitation, as follows from Eq. (2.20).

8.2 Strong Quadrupole Coupling

For the single-pulse experiment the intensity of the central transition, $G_{-1/2,1/2}$, depends on the rf amplitude, $B_{rf} = \omega_{rf}/\gamma$, the pulse duration τ, and the spin I. For nonselective excitation, cf. Eq. (2.20), we have

$$G^{ns}_{-1/2,1/2} = \frac{3(2I+1)}{8I(I+1)} \sin(\omega_{rf}\tau) \tag{8.05}$$

and for selective excitation, cf. Eq. (2.6),

$$G^{s}_{-1/2,1/2} = \frac{3}{4I(I+1)} \sin\{(I+1/2)\omega_{rf}\tau\}. \tag{8.06}$$

Comparing Eqs. (8.05) and (8.06), it can be seen that the maximum intensity of the central transition for the optimum flip angle is divided by $I + 1/2$ going from nonselective to selective excitation. However, for small flip angles is $\sin\{(I + 1/2)\omega_{rf}\tau\} \approx (I + 1/2)\omega_{rf}\tau$ and both intensities are approximately equal. For any spin, any ratio ω_Q/ω_{rf} and a relative deviation of less than 10%, $\frac{G^{ns} - G^{s}}{G^{ns}} < 0.1$, an upper limit for $\omega_{rf}\tau$ can be given [52]: By expanding Eqs. (8.05) and (8.06) one finds

$$\omega_{rf}\tau < \sqrt{\frac{0.6}{I(I+1) - 3/4}} \approx \frac{\pi}{4(I+1/2)}. \tag{8.07}$$

As long as Eq. (8.07) is fulfilled, the observed NMR intensity does not depend on the quadrupole interaction, and, the measured intensities can be compared directly. However, it should be emphasized that the dependencies given by the Eqs. (8.05) and (8.06) describe the intensities of the *central transition*, i.e., it is assumed that contributions from the satellites within the spectral range of the central transition can be neglected.

Spin-echo experiments are necessary for powders if $\nu_Q^2[I(I+1) - 3/4] \gtrsim 2\nu_L/10\,\mu s$. The spin-echo method can also be very useful if two species have a different spin-echo decay, but cannot be resolved by other means.

As discussed in Sect. 5, only a selective excitation of the central transition allows simple echo intensity measurements. The rf inhomogeneity must be kept at a minimum. To go from partly selective excitation into selective excitation, the rf amplitude, ω_{rf}, can be lowered, probes with a higher quality factor improve the selectivity. The actual rf amplitude can be determined either from the selective nutation experiment or from a small liquid standard sample.

If more than one species is present in the sample, the conditions of selectivity, which are a consequence of the quadrupole coupling, may be different for each species. Also, in some cases other line broadening effects may require a larger rf amplitude, so that a selective excitation cannot be achieved.

The resonance offset should be minimized for selective excitation, in order to obtain uniform excitation of the central transition. If a resonance offset due to a second order frequency distribution of the central transition cannot be avoided, the selective π pulse in the echo sequence refocuses small resonance offsets for the central transition. For a partly selective excitation, severe oscillations in the spin-echo decay may be observed due to the resonance offset: The first pulse of, e.g., an in-phase $\pi/2|_x - \pi|_x$ spin-echo sequence creates a transverse magnetization (FID) which is partly refocussed by the second π-pulse and appears with the phase $-y$ at a time Δ after the second pulse. Due to the resonance offset, at a time Δ after the second pulse, the second component of the FID which is created by the second pulse due to spins still in z-direction after the first pulse has the phase $+y + \omega_{off}\Delta$. Hence, for the time where it is superimposed on the spin-echo it causes oscillations of the spin-echo amplitude, which can be removed by phase alternating [303] or cycling [124] of the second pulse. With an increase in the selectivity of the excitation, the intensity of the spin-echo decay approaches the selective one-pulse FID intensity.

Quantitative excitation by a frequency-stepped adiabatic half-passage was introduced by Kentgens [258] on the basis of an approach of Sindorf and Bartuska [259], who changed the technique for the measurement of wide lines with low-power pulses which cover only a small fraction of the linewidth. The magnetic field or the carrier frequency of the pulse is swept. Kennedy et al. [260] used the variable (stepped) frequency pulse method for the study of ^{2}H and ^{27}Al NMR signals and concluded that this method as applied to quadrupole nuclei suffers from signal-to-noise ratio and freedom from artifacts in comparison with the high-power Fourier transform method. Kentgens [258] implemented an adiabatic half-passage by fast phase-continuous frequency switching and modulation [261]. Ignoring relaxation effects, the response to such a half-passage, which varies the effective field from parallel to perpendicular with respect to the external field, is identical to a 90° pulse independent of the ratio ω_{rf}/ω_Q. Kentgens [258] claimed that this method has an advantage over the high-power Fourier transform method, because the latter is limited to a short-pulse excitation in order to obtain true intensities, and demonstrated it with ^{23}Na NMR of a mixture of NaCl and Na_2SO_4. The claimed gain in the signal-to-noise compared with selective excitation lacks a discussion of the corresponding spin–lattice relaxation, which is typically smaller when only the central transition is excited.

8.3 Quantification of MAS Spectra

The application of MAS for the determination of intensities is not recommended, however, it can often be necessary in order to distinguish between different species. MAS does not affect the intensity, G(0), as long as the duration of the pulse is short compared with the rotation cycle. But the spectral distribution of intensity changes and spinning sidebands have to be considered [132]. Some complications when using MAS for intensity measurements of quadrupole nuclei should be briefly mentioned here:

(1) For a small quadrupole coupling an amount of intensity from the satellite transitions appears in the center band (zero-order side band of the satellite transitions). For strong quadrupole coupling (the quadrupole frequency exceeds the rotation frequency by more than two orders of magnitude), this contribution can be neglected. But for $\nu_Q^2 > \nu_L \nu_{rot}/2$ the signal of the central transition also splits into a center band an various spinning sidebands, and, either the sidebands must be added to the center band or the amount of the central transition intensity in the center band must be calculated, cf. Sect. 6.1. The addition procedure decreases the signal to noise ratio and multiplies the contribution from the satellite transition.

(2) Since MAS may average the EFG, the spin–lattice relaxation can change [262] in either direction, and depends on the spinning speed. Therefore, the spin–lattice relaxation has to be determined in the MAS experiment.

(3) The filling factor of the coil as well as the rf inhomogeneity depends on the filling of the MAS rotor. The hole in the middle of the MAS rotor, created by the radial forces, can be different for each sample.

(4) The application of a spin-echo under MAS is very complicated and no comprehensive investigation is known for quadrupole nuclei. The common dead time problem must be solved by baseline fitting, or by starting the acquisition after one rotation period. Taulelle et al. [263] adapted the baseline cosine fitting from Heur and Haberlen [264] to MAS NMR spectra of partly selective excited quadrupole nuclei.

9 Survey of Quadrupole Parameters and Isotropic Values of the Chemical Shift for Some Selected Compounds

A survey of nuclear quadrupole frequency data published up to the end of 1982 is given by H. Chihara and N. Nakamura in Landolt-Börnstein, volume 20 [16]. Values of the chemical shift of quadrupole nuclei in solids can be found in books like "Multinuclear NMR" (edited by J. Mason) [17]. Tables 9.1–9.3 present some electric field gradient and chemical shift data which were obtained by means of NMR techniques and published from 1983 to 1992 for the quadrupole nuclei with half-integer spins ^{27}Al, ^{23}Na, and ^{17}O. For other nuclei we will give only one recent reference for each of them (see also references therein): Lithium-7, Nielsen et al. [137]; boron-11, Bray [283]; sulfur-33, Bastow and Stuart [284]; potassium-39, Bastow et al. [285]; calcium-43, Bryant et al. [186]; scandium-45, Thompson et al. [286]; titanium-47/49, Forbes et al. [287]; vanadium-51, Skibsted [65]; cobalt-59, Hirschinger et al. [288] (see also Eguchi et al. [289]); zinc-67, Bastow and Stuart [284]; gallium-69/71, Timken and Oldfield [290]; rubidium-85/87, Cheng et al. [63] (see also Baltisberger et al. [291]); zirkonium-91, Bastow et al. [292]; niobium-93, Douglass and Peterson [293]; indium-115, Han et al. [167]; antimony-121,

Table 9.1. Quadrupole coupling constant $C_{qcc} = e^2qQ/h$, the asymmetry parameter η, and the isotropic value of the chemical shift δ (refered to 1.0 *M* $AlCl_3 6H_2O$) for some ^{27}Al powder compounds compiled by Dirk Müller [265]. The table contains data published since 1983 and measured at ambient temperature. An asterisk after the reference denotes that also references therein are of importance

Compound	site	C_{qcc}/MHz	η	δ/ppm	Ref.
aluminates					
$CaO 6Al_2O_3$	AlO_4	2.0	$0_{assumed}$	65	[266]
	AlO_5	6.7	$0_{assumed}$	27.5	[87]
	AlO_6-I	1.5	$0_{assumed}$	9	[266]
	AlO_6-II	<1	$0_{assumed}$	16	[266]
$CaO 2Al_2O_3$	AlO_4-I	6.7	0.8	78	[266]
	AlO_4-II	13	0.1	ca. 60	[266]
$CaO Al_2O_3$	AlO_4-I	2.7	0.85	80	[266]
	AlO_4-II	2.7	0.85	83	[266]
$4CaO 3Al_2O_3$	AlO_4	2.4	0.95	80	[266]
$12CaO 7Al_2O_3$	AlO_4-I	3.7	0.9	79	[266]
	AlO_4-II	11	0.2	85	[267]
$3CaO Al_2O_3$	AlO_4	9.7	0.3	85	[267]
	AlO_4-I	8.69	0.32	79.5	[170]
	AlO_4-II	9.3	0.54	78.25	[170]
$3CaO Al_2O_3 6H_2O$	AlO_6	0.69	$0.0_{assumed}$	—	[157]
				12.6	[268]
$4CaO 3Al_2O_3 3H_2O$	AlO_4-I	1.8	0.5	78	[266]
	AlO_4-II	5.4	0.45	79	[266]
$KAlO_2$	AlO_4	1.1	0.7	76	[267]
$KAlO_2 0.5H_2O$	AlO_4	5.6	0.0	77	[267]
$KAlO_2 H_2O$	AlO_4	6.5	0.6	83	[267]
$KAlO_2 1.5H_2O$	AlO_4	5.0	0.25	81	[267]
$5BaO Al_2O_3$	AlO_4	2.3	0.8	80	[267]
$BaO Al_2O_3$	AlO_4	2.4	0.4	78	[267]
$\alpha BaO Al_2O_3 2H_2O$	AlO_4-I	3.4	0.5	81	[267]
	AlO_4-II	5.1	0.9	80	[267]
α-$LiAlO_2$	AlO_6	2.8	0.05	16	[269]
β-$LiAlO_2$	AlO_4	1.8	0.55	82	[269]
	AlO_4	1.86	0.56	83.0	[136]
γ-$LiAlO_2$	AlO_4	3.2	0.7	81	[269]
β-$NaAlO_2$	AlO_4	1.4	0.5	80	[267]
α-Al_2O_3	AlO_6	2.38	0.00	16.0	[136]
aluminium chlorides					
$AlCl_3 3Al(OH)_3 6H_2O$	AlO_6	6.9	0.4	7	[265]
$AlCl_3 4Al(OH)_3 7H_2O$	AlO_6	5.7	0.7	3	[265]
aluminium salts					
$AlO_4Al_{12}(OH)_{25}(H_2O)_{11}(SO_4)_3$	AlO_6	10	0.2	—	[151]
				10	[268]
$Al_2(OH)_2(H_2O)_8(SO_4)_2 2H_2O$	AlO_6	4.6	0.4	3	[265]
$Al_2(OH)_4SO_4 7H_2O$ (aluminite)	AlO_6-I	10.1	0.1	6.9	[265]
	AlO_6-II	11.6	0.15	6.4	[265]
$KAl(SO_4)_2 12H_2O$	AlO_6	0.400	0.00	−4.1	[136]
$NH_4Al(SO_4)_2 12H_2O$	AlO_6	0.456	0.00	−0.4	[136]
alumoborates					
$2SrO Al_2O_3 B_2O_3$	AlO_4	4.3	0.65	83.5	[268]
$2CaO Al_2O_3 B_2O_3$	AlO_4	6.25	0.45	79.5	[268]
$2Li_2O Al_2O_3 B_2O_3$	AlO_4	6.0	0.45	76	[268]
$3Li_2O Al_2O_3 2B_2O_3$	AlO_4	6.7	0.83	70	[268]

Table 9.1. (*Continued*)

Compound	site	C_{qcc}/MHz	η	δ/ppm	Ref.
$9Al_2O_32B_2O_3$	AlO_4	9.1	0.43	70	[297]
	AlO_5-I	7.2	0	52	[297]
	AlO_5-II	9.4	0.49	43	[297]
	AlO_6	6.1	0.44	8	[297]
aluminiumphosphates					
$AlPO_4$ (quartz)	AlO_4	4.2	0.35	44.8	[271]
$AlPO_4$ (tridimite)	AlO_4	0.75	0.95	39.8	[271]
$AlPO_4$ (cristobalite)	AlO_4	1.2	0.75	42.5	[271]
$Al_2PO_4(OH)_3$ (augelite)	AlO_5	5.5	0.78	30	[272]
	AlO_6	4.7	0.2	−3	[272]
$Al_2PO_4(OH)_3H_2O$ (senegalite)	AlO_5	2.87	$0_{assumed}$	30	[272]
	AlO_6	4.09	$0_{assumed}$	1.7	[272]
$KAlP_2O_7$	AlO_6	1.2	0.25	−16	[273]
$AlPO_4$-5 (molecular sieve)	AlO_4	2.3	0.95	40.4	[271]
$AlPO_4$-8 (molecular sieve)	AlO_4-I	3.13	0.50	42.5	[183]
	AlO_4-II	>3.5	—	27	[183]
	AlO_5	>3.5	—	6.5	[183]
	AlO_6-I	2.10	0.50	−9.6	[183]
	AlO_6-II	>3.5	—	18.0	[183]
$AlPO_4$-21 (molecular sieve)	AlO_4	3.7	0.15	48	[169]
	AlO_4	3.7	0.15	47.3	[201]
	AlO_5-I	5.1	0.4	14	[169]
	AlO_5-I	5.9	0.68	14.6	[201]
	AlO_5-II	7.4	0.65	16	[169]
	AlO_5-II	7.4	0.52	15.7	[201]
$AlPO_4$-25 (molecular sieve)	AlO_4-I	1.9	$0.67_{assumed}$	40.8	[201]
	AlO_4-II	0.8	$0.67_{assumed}$	39.5	[201]
$AlPO_4$-25 (dehydrated sieve)	AlO_4-I	2.3	$0.67_{assumed}$	39.2	[201]
	AlO_4-II	1.1	$0.67_{assumed}$	37.5	[201]
alumosilicates					
Al_2SiO_5 (sillimanite)	AlO_4	6.77	0.53	64.5	[156]*
	AlO_6	8.93	0.46	4.0	[156]*
Al_2SiO_5 (andalusite)	AlO_5	5.73	0.7	35	[171]
	AlO_5	5.9	0.7	36	[156]*
	AlO_6	15.5	0.0	10	[171]
	AlO_6	15.7	0.08	—	[156]
Al_2SiO_5 (kyanite)	AlO_6-I	6.53	0.59	7.5	[171]
	AlO_6-I	6.53	0.59	7.5	[156]*
	AlO_6-II	3.7	0.89	5.0	[171]
	AlO_6-II	3.7	0.89	5.0	[156]*
	AlO_6-III	9.37	0.38	13	[171]
	AlO_6-III	9.37	0.38	—	[156]*
	AlO_6-IV	10.03	0.27	15	[171]
	AlO_6-IV	10.04	0.27	—	[156]*
pennine	AlO_4	2.8	—	72	[156]
	AlO_6	1.4	—	10	[156]
$KAl_2[(OH,F)_2/AlSi_3O_{10}]$ (muscovite)	AlO_4	2.1	—	72	[156]
	AlO_6	2.2	—	5	[156]
$CaAl_2[(OH)_2/Al_2Si_2O_{10}]$ (margarite)	AlO_4	4.2	—	76	[156]
	AlO_6	6.3	—	11	[156]
xantophyllite	AlO_4	2.8	—	76	[156]
	AlO_6	2.0	—	11	[156]

Table 9.1. (*Continued*)

Compound	site	C_{qcc}/MHz	η	δ/ppm	Ref.
$Na[AlSi_3O_8]$ (low albite)	AlO_4	3.29	0.62	62.7	[274]
	AlO_4	3.32	0.64	62.6	[262]
$K[AlSi_3O_8]$ (microcline)	AlO_4	3.22	0.21	58.5	[274]
pyrophyllite dehydroxylate	AlO_5	10.5	0.60	29	[275]
natrolite	AlO_4	1.67	0.50	64	[156]*
LiCl-sodalite	AlO_4	0.98	0.59	71.9	[137]
LiBr-sodalite	AlO_4	0.71	0.61	70.9	[137]
NaCl-sodalite	AlO_4	0.94	0.32	62.9	[137]
NaBr-sodalite	AlO_4	0.81	0.29	61.8	[137]
NaI-sodalite	AlO_4	0.57	0.34	60.4	[137]
hydrated zeolite Na-A	AlO_4	1.1	0.75	59.2	[156]*
hydrated zeolite Na-Y	AlO_4	2.0	0.5	62.8	[156]*
dehydrated zeolite Na-Y	AlO_4	7.0	0.3	ca. 60	[270]
dehydrated zeolite Na-ZSM-5	AlO_4	6.1	0.3	ca. 60	[270]
dehydrated zeolite H-Y	AlOHSi	17.1	0.6	ca. 60	[270]
dehydrated zeolite H-ZSM-5	AlOHSi	17.5	0.3	ca. 60	[270]
aluminiumdigermanates					
$Al_2Ge_2O_7$	AlO_5	8.8	0.4	36	[276]
$AlLaGe_2O_7$	AlO_5	7.2	0.37	36	[276]
fluoroaluminates					
$H_3AlF_66H_2O$	AlF_6	0.3	0.0	−2.8	[277]
$K_2AlF_5H_2O$	AlF_6	12	0.0	0	[277]
$Rb_2AlF_5H_2O$	AlF_6	13	0.0	0	[277]
$CsAlF_5H_2O$	AlF_6	7.5	0.15	−10	[277]
NH_4AlF_4	AlF_6	10	0.1	−6	[277]
$KAlF_4$	AlF_6	12	0.0	−5	[277]
$RbAlF_4$	AlF_6	13	0.1	−4	[277]
$Al_2SiO_4F_2$ (topas)	AlF_6	1.7	0.4	0.3	[277]

Table 9.2. Quadrupole coupling constant $C_{qcc} = e^2qQ/h$, the asymmetry parameter η, and the isotropic value of the chemical shift δ (refered to 1.0 *M* NaCl) for the ^{23}Na NMR of some powdered substances at ambient temperature. An asterisk denotes that recent results obtained by DOR NMR were presented in Ref. [207]

Compound	site	C_{qcc}/MHz	η	δ/ppm	Ref.
sodium-nitrogen compounds					
$NaNO_3$		0.337	0.00	−8.0	[136]
$NaNO_2$		1.09	0.11	−8.1	[136]
NaN_3		0.297	0.12	−3.8	[136]
sodalites					
NaCl-sodalite, dehydrated		0–0.5	—	−8.8	[137]*
		0.6–0.8	—	—	[278]
NaCl-sodalite, hydrated		$\leqq$0.1	—	—	[278]*
NaBr-sodalite, dehydrated		0.72	0.12	−9.9	[137]*
		0.7–0.9	—	—	[278]
NaBr-sodalite, hydrated		0.6–0.8	—	—	[278]*
NaI-sodalite, dehydrated		1.73	0.06	−20.6	[137]*
		1.6–1.9	0	—	[278]
NaI-sodalite, hydrated		1.5–1.8	0	—	[278]*
Na-hydroxosodalite		2.00	0.10	3.2	[70]
Na-nitride sodalite		1.00	0.18	0.4	[70]

Table 9.2. (*Continued.*)

Compound	site	C_{qcc}/MHz	η	δ/ppm	Ref.
silicates					
$Na_2H_2SiO_4 8H_2O$		1.11	0.72	3.8	[70]
$Na_2H_2SiO_4 5H_2O$	I	1.35	0.45	5.7	[70]
	II	2.01	0.70	0.0	[70]
$Na_2H_2SiO_4 4H_2O$	I	1.80	0.75	9.0	[70]
	II	2.83	0.17	9.5	[70]
others					
NaOH		3.5	0.00	19.4	[70]
$Na_3P_3O_9$	I	1.57	0.55	1.6	[70]
	II	2.20	0.70	−7.6	[70]
$Na_2(OOCCH)_2 H_2O$	I	1.34	0.80	−2.2	[70]
	II	0.80	0.75	0.9	[70]
$Na_2C_2O_4$		2.50	0.74	17.9	[136]
$CH_3COONa 3H_2O$		0.779	0.38	2.1	[136]
$NaAlSi_3SiO_4$ (low albite)		2.58	0.25	−6.8	[274]
Na_2SO_4		2.60	0.58	−1.3	[70]
		2.4	0.65	—	[168]
Na_2MoO_4		2.45	0.0	—	[150]

Table 9.3. Quadrupole coupling constant $C_{qcc} = e^2qQ/h$, the asymmetry parameter η, and the isotropic value of the chemical shift δ (refered to H_2O) for some ^{17}O powder compounds at ambient temperature. An asterisk denotes that the shift value is confirmed in Ref. [205] and recent results obtained by DAS NMR were presented in Ref. [301]

Compound	site	C_{qcc}/MHz	η	δ/ppm	Ref.
aluminum oxides and hydroxides					
α-Al_2O_3 (corundum)	OAl_4	2.17	0.55	75	[177]
AlO(OH) (boehmite)	OAl_4	1.20	0.1	70	[177]
	OAl_4	1.15	0.13	70.0	[136]
	Al_2OH	5.0	0.5	40	[177]
$Al(OH)_3$ (bayerite)	Al_2OH	6.0	0.3	40	[177]
γ-Al_2O_3	OAl_4	1.8	$0_{assumed}$	73	[177]
η-Al_2O_3	OAl_4	1.6	$0_{assumed}$	73	[177]
δ-Al_2O_3	OAl_4	1.6	$0_{assumed}$	72	[177]
θ-Al_2O_3	OAl_4	1.2	$0_{assumed}$	72	[177]
	OAl_3	4.0	0.6	79	[177]
alumophosphates					
$AlPO_4$-5 (molecular sieve)	AlOP	5.7	0.0	63	[279]
$AlPO_4$-11 (molecular sieve)	AlOP	5.7	0.0	64	[279]
$AlPO_4$-17 (molecular sieve)	AlOP	5.6	0.1	67	[279]
alumosilicates					
Ga-sodalite	SiOGa	4.0	0.3	29	[279]
	SiOSi	5.1	0.0	52	[279]
NaBa-Ga-sodalite	SiOGa	4.0	0.3	29	[279]
	SiOSi	5.1	0.0	52	[279]
zeolite Ga-X	SiOGa	4.0	0.3	28	[279]
	SiOSi	5.0	0.0	50	[279]
zeolite Na-Y	SiOAl	3.1	0.2	31	[280]
	SiOSi	4.6	0.2	46	[280]
zeolite $Ba_{0.65}Na_{0.35}$-Y	SiOAl	3.4	0.2	45	[280]
	SiOSi	5.1	0.1	57	[280]
zeolite NH_4-Y	SiOAl	3.2	0.2	31	[280]
	SiOSi	5.0	0.1	47	[280]

Table 9.3. (*Continued.*)

Compound	site	C_{qcc}/MHz	η	δ/ppm	Ref.
dealuminated zeolite Na-Y	SiOSi	5.2	0.1	47	[280]
zeolite Na-A	SiOAl	3.2	0.2	33	[280]
silicates					
SiO_2 (low cristobalite)	SiOSi	5.3	0.0	46	[280]
	SiOSi	5.3	0.12	40	[302]
SiO_2 (amorphous)	SiOSi	5.8	0.0	50	[176]
	SiOH	4.0	0.3	20	[176]
Mg_2SiO_4 (forsterite)	SiOMg-I	2.35	0.2	61	[281]*
	SiOMg-II	2.35	1.0	62	[281]*
	SiOMg-III	2.70	0.3	47	[281]*
$MgSiO_3$ (clinoenstatite)	SiOMg-I	3.2	0.0	60	[282]*
	SiOMg-II	3.2	0.0	42	[282]*
	SiOSi	5.1	0.3	62	[282]*
$CaMgSi_2O_6$ (diopside)	SiOCa	2.7	0.0	84	[282]*
	SiOMg	2.7	0.1	63	[282]*
	SiOSi	4.4	0.3	69	[282]*
$Mg_3Si_4O_{10}(OH)_2$ (talc)	SiOMg	3.2	0.0	40	[176]
	SiOSi	5.8	0.0	50	[176]
	MgOH	7.3	0.0	0	[176]
$Mg(OH)_2$ polycrystalline	MgOH	6.8	0.0	25	[176]

Kozlova et al. [294]; indium-127, Tarasov et al. [295]; cesium-133, Power et al. [64]; barium-137, Forbes et al. [287]; lanthanum-139, Thompson et al. [286]; lutetium-175, Chu et al. [296].

10 References

1. Abragam A (1961) Principles of nuclear magnetism. Oxford University Press, Oxford
2. Abragam A, Goldman M (1982) Nuclear magnetism: Order and disorder. Clarendon, Oxford
3. Spiess HW (1985) Advances in Polymer Science 66: 23
4. Kanert O, Mehring M (1971) NMR Basic Principles and Progress 3: 1
5. Cohen MH, Reif F (1957) Solid State Phys. 5: 321
6. Sanctuary BC, Halstead TK (1990) Adv. in Magn. and Optic. Reson. 15: 79
7. Zax DB, Bielecki A, Zilm KW, Pines A, Weitekamp DP (1985) J. Chem. Phys. 83: 4877
8. Various authors (1986) Z. Naturforsch. 41a: 1
9. Das TP, Hahn EL (1958) Solid State Phys., Suppl. 1.
10. Jach T (1976) Appl. Phys. Lett. 28: 49
11. Chang J, Connor C, Hahn EL, Huber H, Pines A (1989) J. Magn. Reson. 82: 387
12. Connor C, Chang J, Pines A (1990) J. Chem. Phys. 93: 7639
13. Connor C, Chang J, Pines A (1990) Rev. Sci. Instrum. 61: 1059
14. Fan NQ, Clarke J (1991) Rev. Sci. Instr. 62: 1453
15. Hurlimann MD, Pennington CH, Fan NQ, Clarke J, Pines A, Hahn EL (1992) Phys. Rev. Lett. 69: 684
16. Chihara H, Nakamura N (1988) Nuclear quadrupole resonance spectroscopy data, Landold-Börnstein, New series, Group III, vol 20, Springer, Berlin Heidelberg New York
17. Mason J (1989) (ed) Multinuclear NMR. Plenum Press, New York
18. Sternheimer RM (1950) Phys. Rev. 80: 102
19. Sternheimer RM (1986) Z. Naturforsch. 41a: 24

20. Schmidt PC, Sen KD, Das TP, Weiss A (1980) Phys. Rev. B 22: 4167
21. Sen KD, Schmidt PC, Weiss A (1986) Z. Naturforsch. 41a: 37
22. Watson RE (1958) Phys. Rev. 111: 1108
23. Das TP, Schmidt PC (1986) Z. Naturforsch. 41a: 47
24. Blaha P, Schwarz K, Dederich PH (1988) Phys. Rev. B 37: 2792
25. Lindsay CG, Tossell JA (1991) Phys. Chem. Miner. 18: 191
26. Janes N, Oldfield E (1986) J. Am. Chem. Soc. 108: 5743
27. Ernst RR, Bodenhausen G, Wokaum A (1987) Principles of nuclear magnetic resonance in one and two dimensions. Oxford Univ. Press, London
28. Hatanaka H, Hashi T (1980) Phys. Rev. B 21: 2677
29. Vega S (1981) Phys. Rev. A 23: 3152
30. Kowalewski J, Einarsson L, Nordenskiild L, Rupprecht A (1988) J. Magn. Reson. 76: 337
31. Man PP (1992) Mol. Phys. 76: 1119
32. Furo I, Halle B, Einarsson L (1991) Chem. Phys. Lett. 182: 457
33. Furo I, Halle B (1992) Mol. Phys. 76: 1169
34. Rooney WD, Barbara TM, Springer JCS (1988) J. Am. Chem. Soc. 110: 674
35. Maarel JRCvd (1991) J. Chem. Phys. 94: 4765
36. Nielsen NC, Bildsoe H, Jakobsen HJ (1992) J. Magn. Reson. 97: 149
37. Nielsen NC, Bildsoe H, Jakobsen HJ (1992) Chem. Phys. Lett. 191: 205
38. Jezzard P, Attard JJ, Carpenter TA, Hall LD (1992) Progr. Nucl. Magn. Reson. Spectr. 23: 1
39. Suits BH, White D (1984) Solid State Comm. 50: 291
40. Suits BH, Lutz JL (1989) J. Appl. Phys. 65: 3728
41. Wolf I private commun.
42. Weissbluth M (1978) Atoms and molecules. Academic, New York
43. Zheng ZW, Gan ZH, Sethi NK, Alderman DW, Grant DM (1991) J. Magn. Reson. 95: 509
44. Rose ME (1967) Elementary theory of angular momentum. Wiley, New York
45. Haeberlen U (1976) High resolution NMR in solids, selective averaging. Academic, New York
46. Mehring M (1983) Principles of high resolution NMR in solids. Springer, Berlin Heidelberg New York
47. Blum K (1981) Density matrix theory and applications. Plenum, New York
48. Jeener J (1982) Adv. Magn. Reson. 10: 1
49. Andrew ER, Tunstall DP (1961) Proc. Phys. Soc. 78: 1
50. Schmidt VH (1971) Ampere Summer School 1971 Ljublijana, ed. by Blinc R, J. Stefan Inst. 1972, p. 75
51. Dirac PAM (1958) The principles of quantum mechanics. Clarendon, Oxford
52. Fenzke D, Freude D, Fröhlich T, Haase J (1984) Chem. Phys. Lett. 111: 171
53. Man PP (1986) J. Magn. Reson. 67: 78
54. Jones WH, Graham TP, Barnes RG (1963) Phys. Rev. 132: 1898
55. Stauss GH (1964) J. Chem. Phys. 40: 1988
56. Narita K, Umeda JJ, Kusumoto H (1966) J. Chem. Phys. 44: 2719
57. Baugher JF, Taylor PC, Oja T, Bray PJ (1969) J. Chem. Phys. 50: 4914
58. Taylor PC, Baugher JB, Kriz HM (1975) Chem. Rev. 75: 203
59. Abart J, Palangie E, Socher W, Voitlander J (1983) J. Chem. Phys. 78: 5468
60. Ebert H, Abart J, Voitlander J (1986) J. Phys. F 16: 1287
61. Barnes RG, Torgeson DR, Bray PJ (1988) Phys. Stat. Sol. 147: K175
62. Chu PJ, Gerstein B (1989) J. Chem. Phys. 91: 2081
63. Cheng JT, Edwards JC, Ellis PD (1990) J. Phys. Chem. 94: 553
64. Power WP, Wasylishen RE, Mooibroek S, Pettitt BA, Danchura W (1990) J. Phys. Chem. 94: 5916
65. Skibsted J, Nielsen NC, Bildsoe H, Jacobsen J (1992) Chem. Phys. Lett. 188: 405
66. Alderman DW, Solum MS, Grant DM (1986) J. Chem. Phys. 84: 3717
67. Sethi NK, Alderman DW, Grant DM (1990) Mol. Phys. 71: 217
68. Taulelle F (1990) NATO-ASI Series C 322: 393
69. Müller D (1982) Ann. der Physik 39: 451
70. Engelhardt G, Koller H (1991) Magn. Reson. Chem. 29: 941
71. Freude D, Haase J, Klinowski J, Carpenter TA, Ronikier G (1985) Chem. Phys. Lett. 119: 365
72. Behrens HJ, Schnabel B (1982) Physica 114 B: 185
73. Kambe K, Ollom JF (1956) J. Phys. Soc. Jap. 11: 50
74. Mansfield M (1965) Phys. Rev. 137: A961
75. Haase J, Oldfield E (1993) J. Magn. Reson. A 101: 30

76. Torrey HC (1949) Phys. Rev. 76: 1059
77. Yannoni N, Kahr B, Mislow K (1988) J. Am. Chem. Soc. 110: 6670
78. Li GY, Wu XW (1990) Chem. Phys. Lett. 168: 581
79. Boudot D, Cante D, Brondeau J (1990) J. Magn. Reson. 87: 385
80. Li Gy, Wu XW (1990) Chem. Phys. Lett. 174: 309
81. Samoson A, Lippmaa E (1983) Phys. Rev. B 28: 6567
82. Samoson A, Lippmaa E (1983) Chem. Phys. Lett. 100: 205
83. Harbison GS, Slokenbergs A, Barbara TM (1989) J. Chem. Phys. 90: 5292
84. Harbison GS, Slokenbergs A, Barbara TM (1990) Z. Naturforsch. 45a: 575
85. Kentgens APM, Lemmens JJM, Geurts FMM, Veeman WS (1987) J. Magn. Reson. 71: 62
86. Janssen R, Tijink GAH, Veemann WS (1988) J. Chem. Phys. 88: 518
87. Veeman WS (1992) Z. Naturforsch. 47a: 353
88. Man PP (1990) Chem. Phys. Lett. 168: 227
89. Wokaun A, Ernst RR (1977) J. Chem. Phys. 67: 1752
90. Pandey L, Towta S, Hughes DG (1986) J. Chem. Phys. 85:6923
91. Mijden JAMvd, Janssen R, Veeman WS (1990) Mol. Phys. 69: 53
92. Samoson A, Lippmaa E (1988) J. Magn. Reson. 79: 255
93. Nielsen NC, Bildsoe H, Jackobsen HJ (1992) J. Magn. Reson. 98: 665
94. Haase J, Freude D, Pfeifer H, Lippmaa E, Sarv P (1988) Chem. Phys. Lett. 152: 254
95. Haase J, Pfeifer H (1990) J. Magn. Reson. 86: 217
96. Emshwiller M, Hahn EL, Kaplan D (1960) Phys. Rev. 118: 414
97. Lang DV, Lo DC, Boyce JB, Slichter CP (1974) Phys. Rev. B 9: 3077
98. Lang DV, Boyce JB, Lo DC, Slichter CP (1972) Phys. Rev. Lett. 29: 776
99. Man PP (1990) Mol. Phys. 69: 337
100. Man PP (1991) J. Magn. Reson. 94: 258
101. Hahn EL (1950) Phys. Rev. 80: 580
102. Das TP, Saha AK (1955) Phys. Rev. 98: 516
103. Solomon I (1958) Phys. Rev. 61–65: 110
104. Butterworth J (1965) Proc. Phys. Soc. 86: 297
105. Bonera G, Galimberti M (1966) Solid State Commun. 4: 589
106. Warren WW, Norberg RE (1967) Phys. Rev. 154: 277
107. Weisman ID, Bennett LH (1969) Phys. Rev. 181: 1341
108. Abe H, Yasuoka H, Hirai A (1966) J. Phys. Soc. Jpn. 21: 77
109. Sobral RR, Dasilva XA, Guimaraes AP (1987) J. Magn. Reson. 75: 23
110. Mehring M, Kanert O (1969) Z. Naturforsch. 24a: 768
111. Schoep GK, Valk HJvd, Frijters GAM, Kok HB, Poulis NJ (1974) Physica 77: 449
112. Mansfield P, McLaughlin DE, Butterworth J (1970) J. Phys. C: Solid St. Phys. 3: 1071
113. Halstead TK, Osment PA, Sanctuary BC, Tegenfeldt J, Lowe IJ (1986) J. Magn. Reson. 67: 267
114. Sanctuary BC (1983) Mol. Phys. 48: 1155
115. Lee N, Sanctuary BC, Halstead TK (1992) J. Magn. Reson. 98: 534
116. Compolieti G, Sanctuary GC, Cole HBR (1990) J. Magn. Reson. 88: 457
117. Man PP (1991) Molec. Phys. 72: 321
118. Man PP (1992) J. Chim. Phys. Phys.-Chim. Biol. 89: 335
119. Barata AC, Guimaraes AP (1985) Physica B 130: 484
120. Tsifrinovich VI (1988) Zh. Ehksp. Teor. Fiz. 94: 208
121. Abelyashev GN, Berzhanskij VN, Sergeev NA, Fedotov YV (1988) Phys. Lett. A 133: 263
122. Furo I, Halle B, Wong TC (1988) J. Chem. Phys. 89: 5382
123. Furo I, Halle B, Quist PO, Wong TC (1990) J. Chem. Phys. 94: 2600
124. Kunwar AC, Turner GL, Oldfield E (1986) J. Magn. Reson. 69: 124
125. Dolinsek J (1991) J. Magn. Reson. 92: 312
126. Furo I, Halle B (1992) J. Magn. Reson. 98: 388
127. Furo I (1992) Chem. Phys. Lett. 194: 435
128. Andrew ER (1981) Phil. Trans. R. Soc. Lond. A299: 505
129. Fyfe CA (1983) Solid state NMR for chemists. C.F.C. Press, Guelph
130. Klinowski J (1984) Prog. Nucl. Magn. Reson. Spectrosc. 16: 237
131. Engelhardt G, Buhl JC, Felsche J, Foerster H (1988) Chem. Phys. Lett. 153: 332
132. Maricq MM, Waugh JS (1979) J. Chem. Phys. 70: 3300
133. Herzfeld J, Berger AE (1980) J. Chem. Phys. 73: 6021
134. Frye JS, Maciel GE (1982) J. Magn. Reson. 48: 125
135. Jakobsen HJ, Skibsted J, Bildsie H, Nielsen NC (1989) J. Magn. Reson. 85: 173

136. Skibsted J, Nielsen NC, Bildsöe H, Jacobsen HJ (1991) J. Magn. Reson. 95: 88
137. Nielsen NC, Bildsoe H, Jakobsen HJ, Norby P (1991) Zeolites 11: 622
138. Massiot D, Bessada C, Coutures JP, Taulelle F (1990) J. Magn. Reson. 90: 231
139. Nolle A (1977) Z. Physik A 280: 231
140. Freude D, Behrens H-J (1981) Crystal Research and Technology 16: K36
141. Müller D, Gessner W, Behrens HJ, Scheler G (1981) Chem. Phys. Lett. 79: 59
142. Kundla E, Samoson A, Lippmaa E (1981) Chem. Phys. Lett. 83: 229
143. Ganapathy S, Schramm S, Oldfield E (1982) J. Chem. Phys. 77: 4360
144. Oldfield E, Kinsey RA, Montez B, Ray T, Smith KA (1982) J. Chem. Soc. Chem. Commun. 254
145. Schramm S, Oldfield E (1982) J. Chem. Soc., Chem. Commun. 980
146. Freude D (1985) Adv. Colloid Interface Sci. 23: 21
147. Amoureux JP, Fernandez C, Carpentier L, Chochon E (1992) Phys. Stat. Sol. (a) 132: 461
148. Amoureux JP (1992) Z. Naturforsch. Sect. A 47: 665
149. Lefebvre F, Amoureux JP, Fernandez C, Derouane EG (1987) J. Chem. Phys. 86: 6070
150. Ganapathy S, Shore J, Oldfield E (1990) Chem. Phys. Lett 169: 301
151. Kunwar AC, Thompson AR, Gutowsky HS, Oldfield E (1984) J. Magn. Reson. 60: 467
152. Amoureux JP, Fernandez C, Lefebvre F (1990) Magn. Reson. Chem. 28: 5
153. Samoson A (1985) Chem. Phys. Lett. 119: 29
154. Samoson A, Kundla E, Lippmaa E (1982) J. Magn. Reson. 49: 350
155. Engelhardt G, Michel D (1987) High resolution solid-state NMR of silicates and zeolites. Wiley, London
156. Lippmaa E, Samoson A, Mägi M (1986) J. Am. Chem. Soc. 108: 1730
157. Jacobsen HS, Norby P, Bildsoe H, Jakobsen HJ (1989) Zeolites 9: 491
158. Jäger C (1992) J. Magn. Reson. 99: 353
159. Jäger C, Rocha J, Klinowski J (1992) Chem. Phys. Lett. 188: 208
160. Jäger C, Müller-Warmuth W, Mundus C, Wüllen Lv (1992) J. Noncryst. Solids 149: 209
161. Dixon WT (1982) J. Chem. Phys. 77: 1800
162. Carduner KR (1989) J. Magn. Reson. 81: 312
163. Oldfield E, Timken HK, Montez B, Ramachandran R (1985) Nature 318: 163
164. Mastikhin VM, Zamaraev KI (1987) Z. Phys. Chem. (Neue Folge) 152: 59
165. Dupree R, Holland D (1990): MAS NMR: a new spectroscopic technique for structure determination in glasses and ceramics, in Lewis MH (ed.): New horizon in glass and glass ceramic. Chapman, London, p. 1–40.
166. Eckert H (1992) Prog. Nucl. Magn. Reson. Spectros. 24: 159
167. Han OH, Timken HKC, Oldfield E (1988) J. Chem. Phys. 89: 6046
168. Dec SF, Maciel GE (1990) J. Magn. Reson. 87: 153
169. Alemany LB, Kirker GW (1986) J. Am. Chem. Soc. 108: 6158
170. Skibsted J, Bildsoe H, Jakobsen HJ (1991) J. Magn. Reson. 92: 669
171. Alemany LB, Massiot D, Sheriff BL, Smith ME, Taulelle F (1991) Chem. Phys. Lett. 177: 301
172. Hartmann SR, Hahn EL (1962) Phys. Rev. 128: 2042
173. Pines A, Gibby MG, Waugh JS (1973) J. Chem. Phys. 59: 569
174. Slichter CP (1990) Principles of magnetic resonance. Springer, Berlin Heidelberg New York
175. Woessner DE (1987) Z. Phys. Chem. 152: 51
176. Walter TH, Turner GL, Oldfield E (1988) J. Magn. Reson. 76: 106
177. Walter TH, Oldfield E (1989) J. Phys. Chem. 93: 6744
178. Harris RK, Nesbitt GJ (1988) J. Magn. Reson. 78: 245
179. Blackwell CS, Patton RL (1984) J. Phys. Chem. 88: 6135
180. Morris HD, Ellis PD (1989) J. Am. Chem. Soc. 111: 6045
181. Morris HD, Bank S, Ellis PD (1990) J. Phys. Chem. 94: 3121
182. Kellberg L, Linsten M, Jacobsen HJ (1991) Chem. Phys. Lett. 182: 120
183. Rocha J, Liu X, Klinowski J (1991) Chem. Phys. Lett. 182: 531
184. Rocha J, Klinowski J (1991) J. Chem. Soc., Chem. Commun. 1121
185. Rocha J, Carr SW, Klinowski J (1991) Chem. Phys. Lett. 187: 401
186. Bryant RG, Ganapathy S, Kennedy SD (1987) J. Magn. Reson. 72: 376
187. Edwards JC, Ellis PD (1990) Magn. Reson. Chem. 28: S59
188. Edwards JC, Adams RD, Ellis PD (1990) J. Am. Chem. Soc. 112: 8349
189. Vega AJ (1992) J. Magn. Reson. 96: 50
190. Vega AJ (1992) Solid State Nucl. Magn. Reson. 1: 17
191. Llor A, Virlet J (1988) Chem. Phys. Lett 152: 248
192. Samoson A, Lippmaa E, Pines A (1988) Mol. Phys. 65: 1013

193. Chmelka BF, Mueller KT, Pines A, Stebbins J, Wu Y, Zwangizer JW (1989) Nature 339: 42
194. Mueller KT, Sun BQ, Chingas GC, Zwanziger JW, Terao T, Pines A (1990) J. Magn. Reson. 86: 470
195. Mueller KT, Wooten EW, Pines A (1991) J. Magn. Reson. 92: 620
196. Grandinetti PJ, Lee YK, Baltisberger JH, Sun BQ, Pines A (1992) J. Magn. Reson. (in print)
197. Samoson A, Lippmaa E (1989) J. Magn. Reson. 84: 410
198. Samoson A, Pines A (1989) Rev. Sci. Instr. 60: 3239
199. Wu Y, Sun BQ, Pines A, Samoson A, Lippmaa E (1990) J. Magn. Reson. 89: 297
200. Wu Y, Chmelka B, Pines A, Davis ME, Grobet PJ, Jacobs PA (1990) Nature 346: 550
201. Jelinek R, Chmelka BF, Wu Y, Grandinetti PJ, Pines A, Barrie PJ, Klinowski J (1991) J. Am. Chem. Soc. 113: 4097
202. Chmelka BF, Wu Y, Jelinek R, Davis ME, Pines A (1991) Stud. Surf. Sci. Catal. 69: 435
203. Grobet PJ, Samoson A, Geerts H, Martens JA, Jacobs PA (1991) J. Phys. Chem. 95: 9620
204. Jelinek R, Chmelka BF, Wu Y, Davis ME, Ulan JG, Gronsky R, Pines A (1992) Catal. Lett. 15: 65
205. Mueller KT, Wu Y, Chmelka BF, Stebbins J, Pines A (1991) J. Am. Chem. Soc. 113: 32
206. Müller D, Smith M, Wunsch L (1990) Bruker-Report I
207. Jelinek R, Chmelka BF, Stein A, Ozin GA (1992) J. Phys. Chem. 96: 6744
208. Jelinek R, Ozkar S, Ozin GA (1992) J. Phys. Chem. 96: 5949
209. Opella SJ, Frey M, Gross TA (1979) J. Am. Chem. Soc. 101: 5856
210. Olivieri A, Frydman L, Graselli M, Diaz L (1988) Magn. Reson. Chem. 26: 615
211. Menger EM, Veeman WS (1982) J. Magn. Reson. 46: 257
212. Olivieri AC (1989) J. Magn. Reson. 81: 201
213. Chu P-J, Lunsford JH, Zalewski DJ (1990) J. Magn. Reson. 87: 68
214. Gobetto R, Harris RK, Apperley DC (1992) J. Magn. Reson. 96: 119
215. Böhm J, Fenzke D, Pfeifer H (1983) J. Magn. Reson. 55: 197
216. Komoroski RA, Parker RG, Mazany AM (1987) J. Magn. Reson. 73: 389
217. Harris RK (1988) J. Magn. Reson 78: 389
218. Apperley DC, Haiping B, Harris RK (1989) Molec. Phys. 68: 1277
219. Sastry DL, Naito A, McDowell CA (1988) Chem. Phys. Lett. 146: 422
220. Olivieri A (1992) J. Am. Chem. Soc. 114: 5758
221. Olivieri AC, Frydman L, Diaz LE (1987) J. Magn. Reson. 75: 50
222. Fenzke D, Brunner E publication in preparation
223. Freude D, Klinowski J, Hamadan H (1988) Chem. Phys. Lett. 149: 355
224. Khutsishvili GR (1970) Progress in Low Temperature Physics 9: 375
225. Korringa J (1950) Physica 16: 601
226. Mitchell AH (1956) J. Chem. Phys. 26: 1714
227. Pound RV (1950) Phys. Rev. 79: 685
228. Andrew ER, Swanson KM (1960) Proc. Phys. Soc. 75: 582
229. Kranendonk JV (1954) Physica 20: 781
230. Yosida K, Moriya T (1956) J. Phys. Soc. Japan 11: 33
231. Kondo J, Yamashita J (1959) J. Phys. Soc. Solids 10: 245
232. Kochelayev BI (1959) J. Exp. Theor. Phys. 37: 342
233. Mieher RL (1960) Phys. Rev. Lett. 4: 57
234. Wikner EG, Blumberg WE, Hahn EL (1960) Phys. Rev. 118: 631
235. Tewari DP, Verma GS (1963) Phys. Rev. 129: 1975
236. Hughes DG (1966) Proc. Phys. Soc. 87: 953
237. Kranendonk J, Walker MB (1968) Can. J. Phys. 46: 2441
238. Hughes DG, Reed K (1971) J. Phys. C.: Solid St. Phys. 4: 2945
239. Hikita T, Tatsuzaki I (1975) J. Phys. Soc. Jap. 38: 1083
240. Gordon MI, Hoch MJR (1978) 11: 783
241. Rubinstein M, Taylor PC (1974) Phys. Rev. B 9: 4258
242. Szeftel J, Alloul H (1975) Phys. Rev. Lett. 34: 657
243. Rubinstein M, Resing HA, Reinecke TL, Ngai KL (1975) Phys. Rev. Lett. 34: 1444
244. Reinecke TL, Ngai KL (1975) Phys. Rev. B 12: 3476
245. Szeftel J, Alloul H (1978) J. Non-Cryst. Solids 29: 253
246. Avogadro A, Tabak F, Corti M, Borsa F (1990) Phys. Rev. B 41: 6137
247. Bjorkstam JL, Villa M (1980) Phys. Rev. B 22: 5025
248. Villa M, Bjorkstam JL (1980) Phys. Rev. B 22: 5033
249. Greenbaum SG, Strom U, Rubinstein M (1982) Phys. Rev. B 26: 5226

250. Greenbaum SG, Strom U (1983) Solid State Comm. 46: 437
251. Becker KD (1982) Z. Naturforsch. 37a: 697
252. Tokuhiro T, Iton LE, Peterson EM (1983) J. Chem. Phys. 78: 7473
253. Haase J, Pfeifer H, Oehme W, Klinowski J (1988) Chem. Phys. Lett. 150: 189
254. Haase J. Pfeifer H, Oehme W, Klinowski J (1988) J. Chem. Soc., Chem. Commun. 1142
255. Towta S, Hughes DG (1990) J. Phys.: Cond. Matter 2: 2021
256. Haase J, Park KD, Guo K, Timken HKC, Oldfield E (1991) J. Phys. Chem. 95: 6996
257. Haase J, Oldfield E, Schmitt K (1992) Chem. Phys. Lett. 193: 274
258. Kentgens APM (1991) J. Magn. Reson. 95: 619
259. Sindorf DW, Bartuska VJ (1989) J. Magn. Reson. 85: 581
260. Kennedy MA, Vold RL, Vold RR (1991) J. Magn. Reson. 92: 320
261. Os JWMv, Steeg GWJ, Veeman WS (1991) Rev. Sci. Instr. 62: 1285
262. Woessner DE, Timken HKC (1990) J. Magn. Reson. 90: 411
263. Taulelle F, Bessada C, Massiot D (1992) J. Chim. Phys. Phys.-Chim. Biol. 89: 379
264. Heur A, Haeberlen U (1989) J. Magn. Reson. 85: 79
265. Müller D, private commun.
266. Müller D, Gessner W, Samoson A, Lippmaa E, Scheler G (1986) Polyhedron 5: 779
267. Müller D, Gessner W, Samoson A, Lippmaa E, Scheler G (1986) J. Chem. Soc. Dalton Trans. 1277
268. Müller D, Hübert T, Gessner W, Kentgens APM publication in preparation
269. Müller D, Gessner W, Scheler G (1984) Polyhedron 2: 1195
270. Ernst H, Freude D, Wolf I publication in preparation
271. Müller D, Jahn E, Ladwig G, Haubenreisser U (1984) Chem. Phys. Lett. 109: 332
272. Bleam WF, Dec SF, Frye JS (1989) Phys. Chem. Minerals 16: 817
273. Müller D, Grunze I, Hallas E, Ladwig G (1983) Z. Anor. Allg. Chem. 500: 80
274. Kirkpatrick RJ, Kinsey RA, Smith KA, Henderson DM, Oldfield E (1985) Am. Mineral. 70: 106
275. Fitzgerald JJ, Dec SF, Hamza AI (1989) Am. Mineral. 74: 1405
276. Massiot D, Kahn-Harari A, Michel D, Müller D, Taulelle F (1990) Magn. Reson. Chem. 28: 82
277. Müller D, Bentrup U (1989) Z. Anorg. Allg. Chem. 575: 17
278. Janssen R, Tijink GAH, Veeman WS, Maesen TLM, Lent JFv (1989) J. Phys. Chem. 93: 899
279. Timken HKC, James N, Turner GL, Lambert SL, Welsh LB, Oldfield E (1986) J. Am. Chem. Soc. 108: 7236
280. Timken HKC, Turner GL, Gilson JP, Welsh LB, Oldfield E (1986) J. Am. Chem. Soc. 108: 7231
281. Schramm S, Oldfield E (1984) J. Am. Chem. Soc. 106: 2502
282. Timken HKG, Schramm SE, Kirkpatrick RJ, Oldfield E (1987) J. Phys. Chem. 19: 1054
283. Bray PJ (1987) J. Non-Cryst. Solids 95/96 P. II: 45
284. Bastow TJ, Stuart SN (1988) Phys. Stat. Sol. (b) 145: 719
285. Bastow TJ, Segel SL, Jeffrey KR (1991) Solid State Commun. 78: 565
286. Thompson AR, Oldfield E (1987) J. Chem. Soc., Chem. Commun. 27
287. Forbes CE, Hammond WB, Cipollini NE, Lynch JF (1987) J. Chem. Soc., Chem. Commun. 433
288. Hirschinger J, Granger P, Rose J (1992) J. Phys. Chem. 96: 4815
289. Eguchi T, Nakayama H, Ohki H, Takeda S, Nakamura N, Kernaghan S, Heaton BT (1992) J. Organometal Chem. 428: 207
290. Timken HKC, Oldfield E (1987) J. Am. Chem. Soc. 109: 7669
291. Baltisberger JH, Gann SL, Wooten EW, Chang TH, Mueller KT, Pines A (1992) J. Am. Chem. Soc. 114: 7489
292. Bastow TJ, Smith ME, Stuart SN (1992) Chem. Phys. Lett. 191: 125
293. Douglass DC, Peterson GE (1986) J. Am. Ceram. Soc. 69: 48
294. Kozlova SG, Kriger YG, Goncharuk VK (1987) Zh. Strukt. Khim. 28: 180
295. Tarasov VP, Kirakosian GA, Buslaev YA, Eichhoff U (1990) Z. Phys. 79: 101
296. Chu PJ, Gerstein BC, Tang HD, Shelton RN (1988) Phys. Rev. B 37: 1796
297. Kunath G, Losso P, Steuernagel S, Schneider H, Jäger C (1992) Solid State Nucl. Magn. Reson. 1: 261
298. Sun BQ, Baltisberger JH, Wu Y, Samoson A, Pines A (1992) Solid State Nucl. Magn. Reson. 1: 267
299. Wu Y, Lewis D, Frye IS, Palmer AR, Wind RA (1992) J. Magn. Reson. 100: 425
300. Harris RK, Olivieri AC (1992) Prog. Nucl. Magn. Reson. Spectros. 24: 435
301. Mueller KT, Baltisberger JH, Wooten EW, Pines A (1992) J. Phys. Chem. Minerals 19: 307
302. Spearing DR, Farnan I, Stebbins JF (1992) Phys. Chem. Minerals 19: 307
303. Man PP (1992) J. Magn. Reson. 100: 157

Solid State NMR of Synthetic Polymers

N. J. Clayden

ICI plc, Wilton Materials Research Centre, P.O. Box No. 90, Wilton, Middlesbrough TS6 8JE, USA

Table of Contents

1 Introduction

Over the past ten years or so solid state NMR has become widely used in the study of polymers. The rather late development of solid state NMR, in comparison to solution state, can be attributed to the need for technical developments to permit the acquisition of high resolution solid state NMR spectra. Since the introduction of the combined cross-polarisation and magic angle spinning technique to polymers by Stejskal and Schaeffer an impressive growth has taken place. It is perhaps appropriate then to take stock of the value of solid state NMR, to point out its strengths and weaknesses, within the context of the polymer industry. With this in mind, my aim in this review is not to provide a comprehensive coverage of the literature but rather to use examples to provide an assessment of the use of NMR in the study of solid materials.

Although it is perhaps stating the obvious, the starting point for any discussion on the value of an analytical technique within the polymer industry must be the reasons for the preference of one particular technique over another. To answer this it is necessary to view the problem from the business perspective; the favoured technique will be the one which can solve the problem quickest and cheapest. In effect the function of the technique is to bring a competitive edge to the business. This is not to say that the sole function of solid state NMR, for example, is to solve the short term problems, background studies which are not directly linked to any business area problem and technique development are also important, but the main concern must be on the former otherwise there may not be a business for you to work in. A preference for a particular technique can be attributed to one or more of three criteria, besides of course a fashion for the newest, these are uniqueness, relative scientific benefit and cost. If only one technique can give you the information then clearly this uniqueness will favour its use. More commonly a number of techniques could be used, for example solid state NMR or FTIR to identify a solid polymer. Here solid state NMR might have the advantage because quantitative data are required. However, if a routine process control measurement is required then FTIR is cheaper and the problem with quantitation is overcome by calibration.

With these criteria in mind the application of solid state NMR can be divided into five areas;

1. Composition
2. Conformational analysis
3. Morphology
4. Orientation
5. Dynamics

Composition and conformational analysis need to be explained in more detail. By composition, I mean the types of structural units present, such as methyl or methylene groups, which can be taken as far as the molecular structure. The area of conformational analysis includes the three-dimensional packing which corresponds to the lattice of the crystalline component of a semi crystalline polymer but is mainly concerned with the molecular conformation. Almost without exception solid state NMR should not be used to study the composition of a soluble polymer, since solution NMR is far better for this. Exceptions to this are usually better phrased as a morphology question, for example in a semi crystalline copolymer are the compositions of the crystalline and amorphous phase different? For insoluble polymers the composition can be studied by FTIR. The other four areas can be regarded as involving specific solid state effects and the competitive technique would be different hence; for conformational analysis, X-ray or neutron diffraction; for morphology, microscopy both optical and electron; for orientation, birefringence, IR dichroism and for polymer dynamics, inelastic neutron scattering.

From this we can see that solid state NMR is often not the only technique available, placing a greater emphasis on the need to show a relative scientific benefit or cost advantage. In terms of the poor availability of alternative techniques the greatest strength of NMR is in the study of molecular motion.

The principal difference between solution and solid state NMR is the resolution in the spectrum. Typically a linewidth of <1 Hz is seen in solution while for a semicrystalline polymer it is of the order of 1–2 ppm for ^{13}C[1] and 0.5 ppm for ^{1}H in the solid [2]. Two principal consequences follow from this; first, ^{1}H NMR is of restricted value given the small chemical shift range of 10 ppm. In fact high resolution ^{1}H NMR of solids is rare. Secondly, structural investigations which depends on small differences in chemical shift, such as sequencing, are in general not possible in the solid state.

A through discussion of the theoretical aspects of solid state NMR can be found in a number of books [3–5], while general applications of solid state NMR are served by a book [6] and reviews [7, 8]. More pertinent descriptions of solid state NMR in the polymer area are provided by the books [9, 10]. In the light of the comprehensive coverage provided by the above sources I reemphasise the aim of this article is to show practical applications within the polymer industry.

2 Composition

An important strength of solid state NMR is the ability to give quantitative data on the composition of an insoluble material whether it be a thermoset or thermoplastic. Three classes of insoluble polymer can be distinguished depending on whether the monomers and cross-linking site are known, Table 1. According to the class of the polymer the type of information we hope to get from the solid state NMR experiments differs. A possible complication in CPMAS NMR is the presence of spinning sidebands which may overlap other resonances of interest, as in Fig. 1. Fast spinning is therefore to be recommended not only from the point of view of resonance overlap but also for sensitivity. As you spin faster so

Table 1. Classes of polymer types

Class	Monomer	Connection	Degree of Connection	Example	Information
K	Known	Known	Known	Polyphenylene sulphide Vinyl resin	Extent of reaction
C	Known	Known	Unknown	Alkoxysilanes	Degree of cross-linking
		Unknown	Known	Phenol-formaldehyde resin	Nature of the cross-link site
		Unknown	Unknown	Alkyne functionalised resin	
U	Unknown	Unknown	Unknown	Plasma polymer	Structural groups

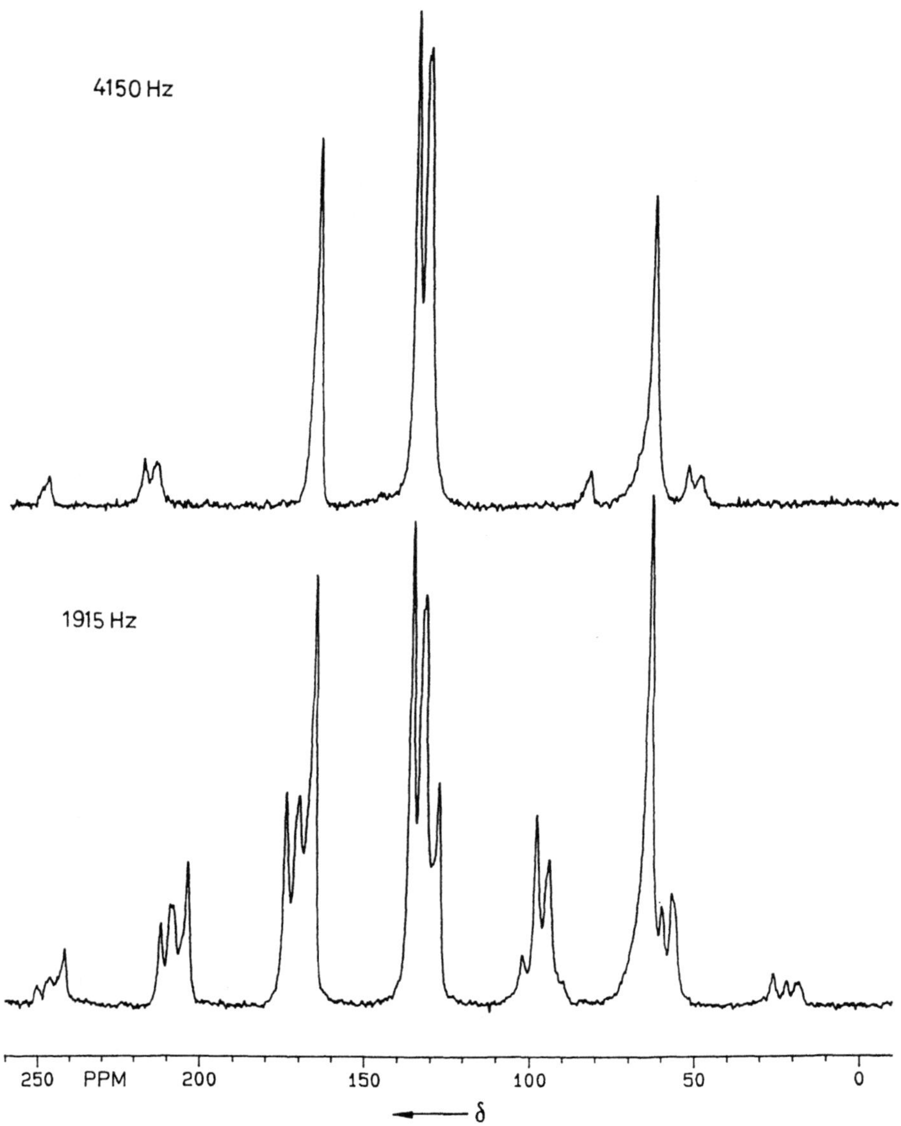

more of the intensity lies in the isotropic resonances. Ideally the first-order spinning sidebands should lie outside the region in which any resonances are expected. Take as an example an aromatic polymer with about 3% of an aliphatic additive; now at 4.7 T, 50.3 MHz for ^{13}C you need to spin at over 7 kHz, equivalent to 140 ppm to ensure that no confusion will occur. Even at this spinning speed the intensity of the spinning sideband will be comparable to the additive you are looking for. At higher field strengths the problem becomes more acute, thus at 75.48 MHz a spinning speed of 10.5 kHz is required. Once this speed is used problems with intensities in a cross-polarisation experiment can emerge [11]. Slower speeds could be used in conjunction with a sideband suppression routine such as TOSS but great care would be necessary to ensure that reliable results are obtained [12].

Slow spinning is used for two reasons, first to allow the chemical shift tensor to be determined through a spinning sideband analysis [13] and secondly to study the molecular orientation when combined in a two dimensional experiment [14]. I shall return to the second point latter. With regard to the first point, to the best of my knowledge the chemical shift tensor has not been used to study the chemistry of polymers but only the molecular dynamics.

2.1 Identification

One of most straightforward uses of solid state NMR is to identify the nature of an insoluble polymer, either by comparison of the NMR spectrum with spectra of standard samples or by assignment of the observed spectrum. In principle NMR can give quantitative results on a sample which consists of a mixture of polymers. As with the main alternative, FTIR, whether you can actually quantitatively identify a mixture depends on achieving a resolution of the components in the spectrum. Quantitation can be a problem with cross polarisation [15, 16] especially in a heterogeneous sample where in the extreme one phase may be too mobile to cross-polarise properly. Under these circumstances a Bloch decay experiment may have to be used. Identifying a polymer in a filled material is as simple as for a polymer alone unless the composite is a long fibre carbon fibre one in which case severe detuning of the probe will be a problem. Usually this type of identification of an unknown polymer would be restricted to the analysis of a competitors product since unless the polymer is of the unknown monomer class the molecular structure will probably be known, though perhaps not the nature of the cross-link site.

Polymers formed by the reaction of alkene functionalised monomers are a good example of where both the monomer and cross link site are known. After the reaction the molecular connectivity is known, not always though as will be described later, and the main consideration is the extent to which the reaction

◄

Fig. 1. Comparison of a relatively slow spinning CPMAS NMR spectrum of poly ethylene terephthalate (PET) and a fast spinning spectrum illustrating the problem of spinning sidebands

has taken place. An illustration of this is provided by the work on poly [oligo (ethylene glycol)dimethacrylates] which involved a series of homologous monomers, $CH_2{=}C(CH_3){-}CO(OCH_2CH_2)_xOCOC(CH_3){-}CO(OCH_2CH_2)_x{\cdot}OCOC(CH_3){=}CH_2$, with x ranging from 1 to a number average of 22 [17]. Two methods were used to estimate the extent of the cure. In the first the cure was obtained from the integrated intensity of all the quaternary carbon resonances relative to the integrated intensity of all the α-methyl resonances in the CP MAS NMR spectrum. By using the carbonyl resonances instead in the second method it was possible to differentiate between constrained unsaturation which was identified with an unsaturated pendant group and unconstrained unsaturation from free monomer. This differentiation was based on the fact highly mobile free monomer resonances will not appear in a cross-polarisation experiment. It was found that the extent of the full cure increased with x, as was predicted, since the network requires a sufficient flexibility to allow all the vinyl groups to react. A valuable aspect of this study, and one which is worth stressing is that a series of samples were studied giving a greater confidence in the interpretation of the curing characteristics.

Given the ability to determine quantitatively the extent of a full cure then clearly solid state NMR can be used to follow a curing process, either for a radical initiated or a thermally initiated reaction providing the reaction time is not too short. The transformation from a liquid to a solid phase is not a problem because with suitable inserts it is possible to spin liquids in an MAS probe at speeds of 2 kHz or so. In fact a greater restriction is the time needed to accumulate the data, for most realistic reactions, with a critical period of ten minutes or less, isotopic enrichment is essential for ^{13}C NMR. Although the early fluid stages of a reaction can be followed by ^{1}H NMR once the matrix effectively gels multiple pulse methods are required to acquire the spectrum and as we have seen the practical value is greatly diminished by the poor resolution.

A more complex reaction is provided by resins where a vinyl functionalised oligomer is present together with a reactive solvent, such as methyl methacrylate. Depending on the relative amounts of the solvent and oligomer a matrix of variable composition could be produced with indeterminate runs of poly methyl methacrylate cross-linking the oligomers. Solid state NMR can be used to determine the cure and to what extent the residual unsaturation is free solvent. However, can it be used to examine the monomer connectivity, in effect to sequence the polymer chain? Sequence information in the NMR spectrum is almost invariably obtained from the small differences in chemical shift which arise from placing a monomer unit between various other units. Usually the differences in chemical shift are quite small and given the broader lines in solid state NMR spectra almost never resolved. Matters are often complicated further by the presence of long runs of one type with only a few different units leading to their low intensity.

Unfortunately it is not possible to say whether this is the case simply from the observed line width. Thus sequencing of the resulting matrix is in general not feasible.

2.2 Unknown Cross-Linking

A more extensive examination of the reaction chemistry is justified when the nature of the cross-linking is unknown. We can consider two types of unknowns, in the first the cross-linking unit is actually known but the degree to which it has reacted is not. Here alkoxysilane based coatings provide an excellent example since the cross-link is known to be silicon but the number of Si–O–Si bonds formed is not. The second type of unknown is where the cross-linking site itself is not known, for example in phenol-formaldehyde resins where the proportion of methylene bridges and the degree of aromatic ring substitution may not be known [18]. Polymerisation involving alkoxysilanes proceed by two steps, first the hydrolysis of the alkoxy group to give a hydroxyl and secondly the formation of an Si–O–Si bond. As the ^{29}Si NMR spectrum of a Degussa 153 based coating illustrates, Fig. 2, differing degrees of Si–O–Si bond formation can be readily seen [19]. However, the first step of hydrolysis is not apparent in this spectrum because the exchange of an alkyl group by a hydrogen causes only a small change in the ^{29}Si chemical shift which cannot be resolved. To some degree this is fortunate because with the broad resonances we might otherwise have to contend with resonances of differing Si–O–Si connectivities overlapping. Although the ^{29}Si NMR spectrum does not help us study the hydrolysis step

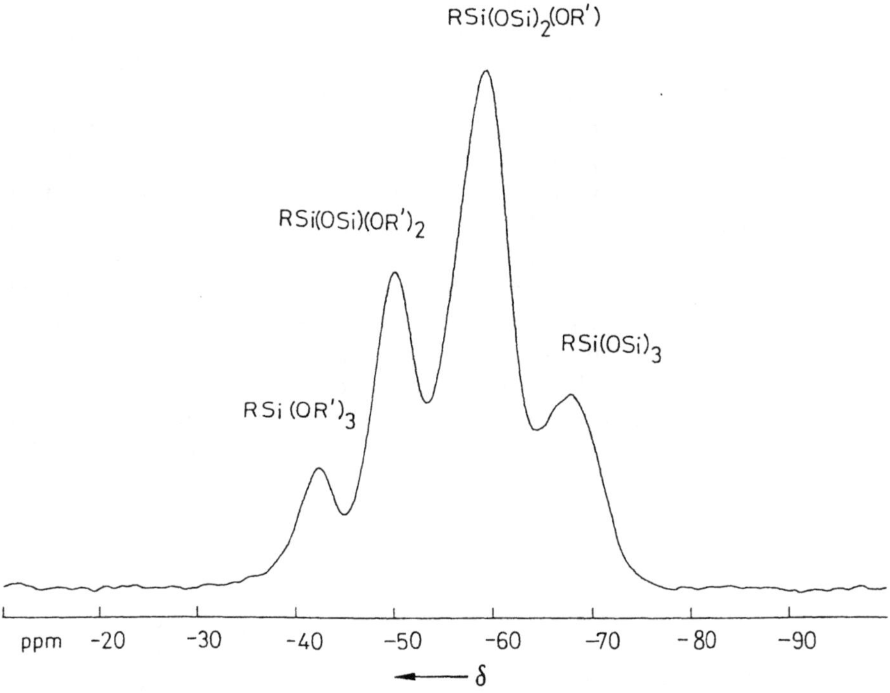

Fig. 2. ^{29}Si CP MAS NMR spectrum of an alkoxysilane coating showing the different kinds of silicon cross-linking unit

some data can be gleaned from the ^{13}C CP MAS NMR spectrum since the α carbon chemical shift is sensitive to the process. In an alkoxysilane the chemical shift of the carbon α to the silicon is in the range 5 to 10 ppm depending on the silicon–silicon connectivity while for the hydrolysed group the range is 7 to 15 ppm [20]. Despite the overlap it is possible to see the overall state of hydrolysis because we know the silicon connectivity from the ^{29}Si NMR spectrum.

Now in order to relate the observed cross-linking to the properties or history of the coating it is essential that quantitative intensities can be measured. Cross-polarisation is essential if we are to acquire the ^{29}Si spectrum in a reasonable time or at all given the low silicon levels in some samples. However, we have noted earlier that cross-polarisation can distort the relative intensities in the spectrum. In view of this it is necessary that a model system, closely related to the one in hand, be taken and the cross-polarisation intensities compared with the fully relaxed single excitation ones. This will require the spin-lattice relaxation time of whatever nucleus is being studied to be measured. For the Degussa 153 coating a T_1 of 82 s was found for all the ^{29}Si resonances and a comparison of the cross-polarisation and single pulse excitation experiments revealed that quantitative results could be obtained from the cross-polarisation spectra [21]. It should be noted that if two or more different alkoxysilanes are present there is no simple way to establish whether a cross-reaction has taken place especially if partial hydrolysis occurs. This is a further illustration of the lower level of structural analysis possible by solid state NMR compared with solution methods largely arising from the greater linewidths. At the same time it must be stressed that NMR is unique in being able to study alkoxysilane cross-linking.

A major problem with the use of solid state NMR to identify a polymer cross-linking site is the empirical nature of the chemical shift. Assignment of the NMR spectrum can be difficult so that the precise type of the cross-linking remains unclear. An example of the problem faced is the chemistry involved with an ethynyl functionalised resin. Polymerisation of the ethynyl group could give a variety of products but one of particular interest is the cyclotrimerisation to a trisubstituted benzene [22]. This reaction can be thermally or catalytically induced. Take the cyclotrimerisation reaction; both the 1,3,5 and 1,2,4 substituted cyclotrimerisation products will have an aromatic quaternary carbon in the range 140–142 ppm while the unsubstituted carbons will appear in the range 125–130 ppm [23]. Uncatalysed curing gives a sample whose spectrum is shown in Fig. 3a. There is no evidence for a cyclotrimerisation product since the

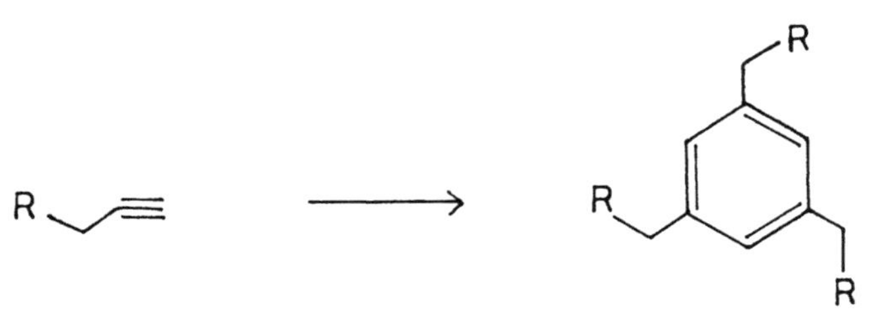

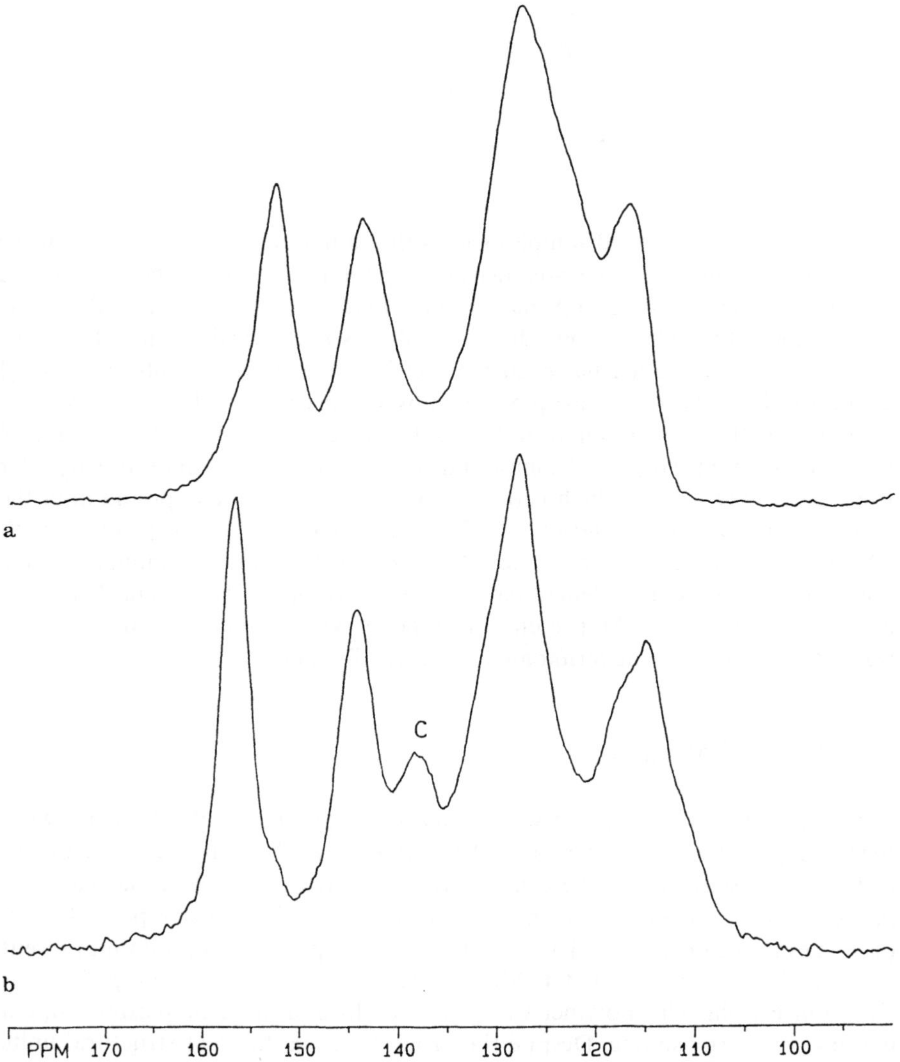

Fig. 3 a, b. ^{13}C CP MAS NMR spectra of a catalysed **b)** and uncatalysed **a)** curing of an ethynyl terminated resin showing evidence for a cyclotrimerisation product, the resonance labelled *C*

resonances at 152 and 143 ppm have an equal integrated intensity as expected from the monomer structure. With a catalyst a quite different spectrum is seen, Fig. 3b, now we have a resonance at 138.2 ppm which a dipolar dephasing spectrum shows to be a quaternary. From the integrated intensity of this resonance we can estimate that no more than 40% of the ethynyl end groups trimerise to benzene during the cross-linking, very similar to the result seen for an ethynyl terminated polyimide resin [22]. Inspection of the spectra from

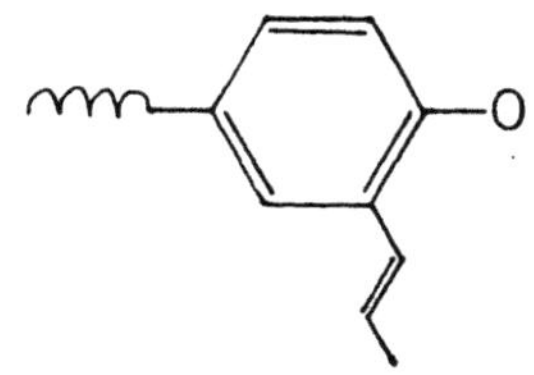

uncatalysed and catalysed samples shows that a major difference occurs in the vinyl carbon region i.e. the resonance at 127 ppm. In the case of the uncatalysed reaction all the ethynyl groups have been converted to vinyl groups while only 60% in the catalysed reaction. One possibility is that substitution at the *ortho* position in the aryl ether has occurred; but here the chemical shift of the vinyl group would be closer to 136 ppm. Again by suggesting a likely product we can examine whether it is present or not. This illustrates too a weakness of the NMR data, in the absence of model compound chemical shifts we cannot directly infer the type of vinyl group which is present. Even with model compound data the broad resonances in the solid state NMR spectrum will restrict the precision with which we can assign the resonance. More often than not a seemingly simple question involves a great deal of background work on model systems before an answer can be given. At present the type of vinyl group produced in the polymerisation of alkyne terminated resins is still unclear.

2.3 Plasma Polymers

In the systems discussed so far we can say with a great deal of confidence what the basic type of structure is because the monomers are known and the chemistry rather circumscribed. On the otherhand the matrices formed in plasma polymerisations are difficult to define. Given the range of chemistry possible in a plasma even the most basic determination of the types of structural unit present is invaluable. A typical ^{13}C CP MAS NMR spectrum of a plasma polymer is shown in Fig. 4a. This polymer was generated from a methane plasma using a microwave energy source. Despite the broad lines six different structural units can be seen either in the standard cross-polarisation or dipolar dephased spectrum, Fig. 4b. The clear advantage over FTIR is that quantitative data can be collected which has been used to determine the "sp^2/sp^3" ratio in various plasma polymers [24, 25]. A good correlation is seen between this ratio and properties of plasma polymers such as hardness and band gap. In passing it should be noted that as the hydrogen content in the polymer decreases so the assumption that all the sample can be cross-polarised becomes increasingly invalid. Hence in low hydrogen content materials it becomes all the more necessary to carry out checks such as acquiring a single pulse excitation spectrum even if this requires the use of isotopic enrichment.

Division of the spectral intensity into "sp^2" and "sp^3" is rather easy because of the clear differences in chemical shifts allowing an accurate integration of the

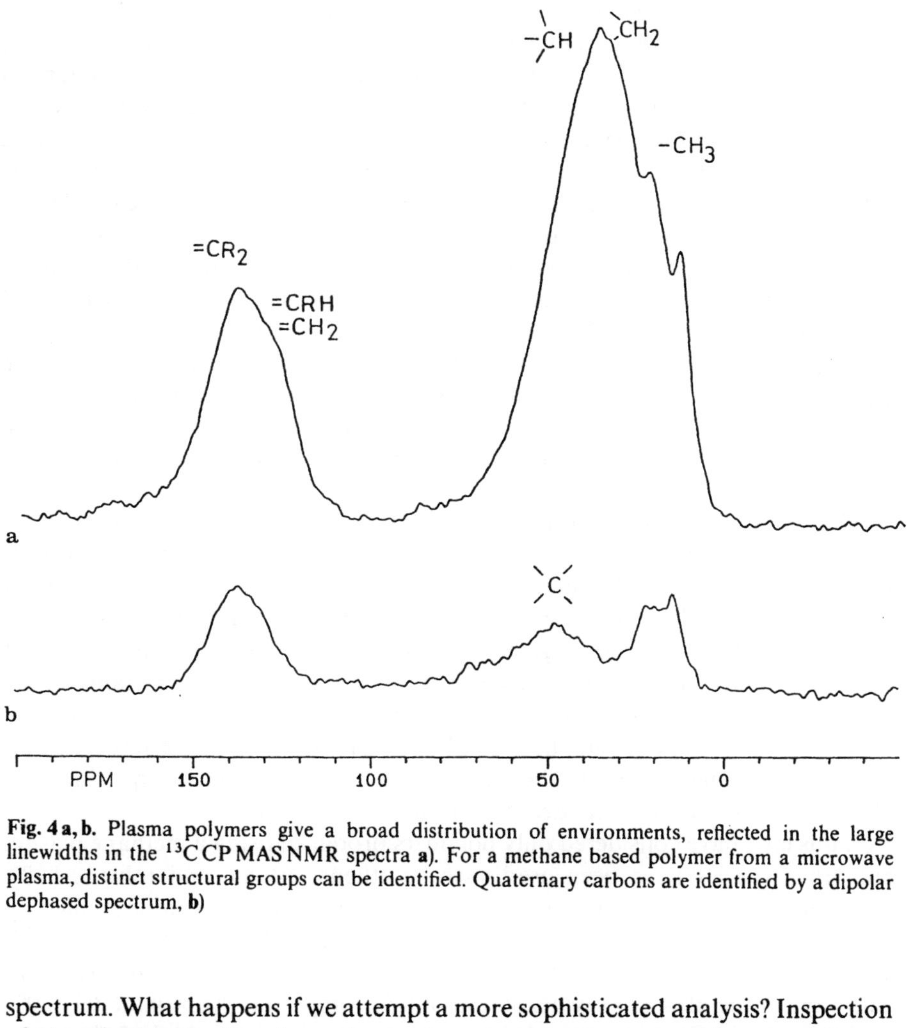

Fig. 4 a, b. Plasma polymers give a broad distribution of environments, reflected in the large linewidths in the ^{13}C CP MAS NMR spectra **a**). For a methane based polymer from a microwave plasma, distinct structural groups can be identified. Quaternary carbons are identified by a dipolar dephased spectrum, **b**)

spectrum. What happens if we attempt a more sophisticated analysis? Inspection of the dipolar dephased spectrum suggests a quaternary carbon at 56 ppm, possibly a true diamond-like structural unit. Can we accurately determine its amount in the plasma polymer? One possibility is to find the time constants in the dipolar dephasing of the resonance intensity at 56 ppm. As expected the decay is two component, a faster second-order decay of the methylene and methine carbons and a slow first-order decay of the quaternary carbon. Comparison of the time constants with literature values suggest they are consistent with the types of group to which they have been assigned [26]. From the dipolar dephasing analysis the amount of the quaternary carbon can be found. However, it was observed that the time constant associated with this carbon was quite variable, to the extent that an accurate measurement of the quaternary carbon content would necessitate determining the full dipolar dephasing behaviour for each sample, clearly a time consuming business [27].

A second possibility is to perform a spectral deconvolution by assuming a functional form for the lineshape and using a non-linear least squares fitting routine. Since we have clear evidence for only six types of structural unit at most six lineshapes were included in the deconvolution. No difficulties were encountered fitting the resonance intensity around 130 ppm using an unconstrained fit, however, for the aliphatic region major errors arose because the underlying resonance at 56 ppm was given an excessive linewidth for example 2000 Hz in a fit compared to ca 1600 Hz experimentally. From a series of plasma polymers it was clear that an unconstrained fit was not appropriate, some constraint on the linewidth was essential. Since all six resonances could be individually resolved in the selective ^{13}C CP MAS NMR experiments accurate constraints on the linewidth and peak position could be found to be used in a constrained fitting procedure. Overall the constrained fits gave a poorer rms error indicating inadequacies in the six resonance approximation but did allow the more accurate deconvolution of the spectrum in terms of giving realistic peak widths and positions.

Although this particular use of spectral deconvolution involves plasma polymers the lessons to be learnt are more widely applicable. First the fit must not be over determined, care must be taken when selecting the number of lineshapes. Secondly when unresolved peaks are included in a deconvolution unconstrained fits are liable to give nonsensical results. This can only be avoided by constraining the peak parameters, but to impose realistic constraints some resolution of the peak is required. In effect an unresolved peak in the standard CP MAS NMR experiment should only be included if it can be resolved in a selective experiment. Despite the sense in these guidelines it is surprising how often over elaborate spectral deconvolution are performed.

Thus far I have considered only polymers produced using a methane plasma but other species have been used such as ethene, ethyne and toluene [28]. As the hydrocarbon becomes more complex so the chemistry can be studied by specific ^{13}C labelling. For example Kaplan and Dilks [28] used selective ^{13}C labelling of the ring and methyl carbons of toluene to follow the reaction pathways in the plasma polymerisation of this compound. Approximately 20% of the aromatic C1 carbons ended up as saturated carbons while 24% of the methyl carbons finished up as unsaturated carbons. Just as specific labelling helps to unravel the plasma reaction mechanism so it can be used to monitor changes in the chemistry as the deposit becomes thicker. In this case ^{13}C depleted gas could be used at the start of the reaction and the ^{13}C enriched gas substituted some time latter.

It is perhaps worth making clear that NMR is not really suitable for looking at the production type plasma polymers which are sub micron in thickness unless the substrate itself is of the same order of thickness. Whilst NMR is fine for studying the reaction chemistry and identifying the structural groups in a model type plasma polymer the poor sensitivity of the technique means that in typical coating applications of plasma polymers the signal is of low intensity and, for organic substrates, will be overwhelmed by the substrate spectrum. These comments are equally valid for a wide range of coatings and films. Model systems are readily studied but in situ examples are often hard.

3 Conformational Analysis

NMR is sensitive to the short range structure around each nuclear spin and through this the overall lattice periodicity in a crystalline material. Although the NMR spectrum of a particular crystal form is generally characteristic and as such can be used as a fingerprint of that structure this is not a physical necessity. As I intend to illustrate the fundamental problem with using the NMR spectrum for structural determinations is the interpretation of the chemical shift. By making use of the dipole–dipole coupling between pairs of spins, such as $^{15}N-^{1}H$ and $^{13}C-^{13}C$, it is possible to measure accurate internuclear distances in specific cases [29, 30]. As a general method for determining the structure of a polymer it is not feasible. Essentially using NMR to derive structural information of a polymer, which is better found by X-ray diffraction, misses the true value of the technique which is to use the NMR spectrum as a fingerprint, especially if the polymer is amorphous. Despite the common assertion that NMR can be used to study the structure of an amorphous polymer the data do not support it. True, NMR can be used to look at the molecular connectivity or rather the types of structural groups present but details of the packing of polymer chains in thermoplastics or the network in a thermoset have not been forthcoming. Similarly for the crystalline phases of a polymer, NMR has not been used to derive the three dimensional structure, rather the NMR spectra have been discussed to explain how they are consistent with the crystal structure. An example of the pitfalls is provided by the work of Tonelli on smectic isotatic polypropylene [31]. The authors present an argument based on the similarity of the ^{13}C NMR chemical shifts and ^{13}C T_1, relaxation times of the β-form and smectic form to conclude that the local packing of chains is similar in the β- and smectic forms. This would imply that the 3_1 helical groups thought to be present in the β form are in groups of the same handedness in the smectic form in contrast to the α-form where they are of opposite handedness and enmeshed. Yet this interpretation would require unwinding and rewinding of the 3_1 helices in the solid–solid smectic to α transition. For this reason they argue that the helices are of opposite handedness in the smectic form but in contrast to the α form are open, not enmeshed. But now we have an open α form, not the β- form to which the NMR data pointed. In short NMR was not used to derive the structure of the smectic form rather it was shown to be consistent with an open α-form chosen for the more compelling reason that it better explains the smectic to α-transition. It is even debatable whether the NMR evidence is consistent with this open α-form rather than the β form. Obviously care must be taken in inferring structural information from NMR data.

Why is it then that although the NMR spectrum is sensitive to the short range structure it is apparently not possible to pull out details of the local environment. The reason is that the dominant factor controlling the chemical shift of a resonance is intramolecular, essentially its nearest neighbours. NMR chemical shifts are indeed a local phenomenon. Interchain effects tend to be rather small and from various sources leading to a generalised line broadening. Disentangling these contributions to the linewidth has not proved to be possible, greatly restricting the structural use of NMR. Even when small splittings are observed as in the case of α isotactic polypropene a precise interpretation of the splitting is difficult, though the spectrum can still be seen to be consistent with the crystal structure. This emphasises that the utility of solid state NMR in the study of semi crystalline polymers must be to use the NMR spectrum as a fingerprint of the crystal structure allowing the effects of processing to studied more readily. Such an empirical use of NMR reflects its strength, the sensitivity of the chemical shift to structural changes and its weakness, the interpretation of the chemical shift in structural terms.

In contrast to the interchain packing effects intra-chain conformational effects tend to be more specific allowing a more detailed investigation of the conformation. An excellent review of conformational analysis in polyalkenes is provided by Ref. [10]. Rather than reiterate the content of this review I will illustrate the type of analysis possible using polypropene as an example. For the polyalkenes the most important interaction is the γ-gauche whereby a *gauche* arrangement leads to a shielding of ca 4.4 ppm relative to a *trans* conformation [32]. This is a specific example of a more general phenomenon namely the effect of the steric interaction between hydrogens on the chemical shift of the carbon to which they are bonded [33]. The effect this can have on the NMR spectrum is well illustrated by the chemical shift data for polypropene, Table 2, in the isotactic and syndiotactic forms. Examination of the crystal structure of syndiotactic polypropene reveals two equally probable methylene carbon sites, one of which has both γ-carbons in the *trans* conformation and the other has two *γ-gauche* carbons accounting for the splitting in the methylene region of 8.7 ppm [34].

On the other hand for isotactic polypropene there is only one methylene carbon environment which has a *trans* and *gauche* γ-carbon giving a single resonance at an intermediate chemical shift. However, it must be noted historically a variety of techniques have been used to study polypropene so it is not surprising that solid state NMR is little used in industry to probe its structural behaviour.

Table 2. Chemical shifts for polypropene

	CH (ppm)	CH_2 (ppm)	CH_3 (ppm)
isotatic	26.5	44.5	22.5
syndiotactic	26.8	39.6, 48.3	21.0

In poly aromatics the *γ-gauche* effect is of little relevance but an extremely common feature is that of splitting of resonances which are ortho to an oxygen substituent, see Fig. 5. Thus inequivalence has been observed in poly phenylene oxide and DGEBA [35]. Small molecular species such as 1,4 diphenoxybenzene also show this splitting which in this case ranges from 0 ppm for the central ring to 6.2 ppm for the terminal ring. Taking into account a number of different shift mechanisms namely, ring current shifts, steric shifts, magnetic anisotropy and electric field shifts an explanation for the inequivalence in ether compounds could be found in the electric field shift originating from the oxygen [36]. Owing to the dominance of the effect at large phenyl ring–oxygen torsion angles there is a direct correlation between this torsion angle and the chemical shift inequivalence.

Despite the expectation of seeing *ortho* carbon inequivalence in polymers such as PEEK, poly ether ether ketone, it has been difficult to find. A first complication is that two types of carbon *ortho* to an ether are present. To overcome this selectively deuterated samples were prepared corresponding to deuteration in the ether–ether and ether–ketone rings respectively. Then to ensure that only those carbons directly bonded to a hydrogen were seen a short contact time of 20 μs was used in the cross polarisation. Lastly in order to pick out only the crystalline phase a delay of 10 ms was used in a delayed contact experiment.

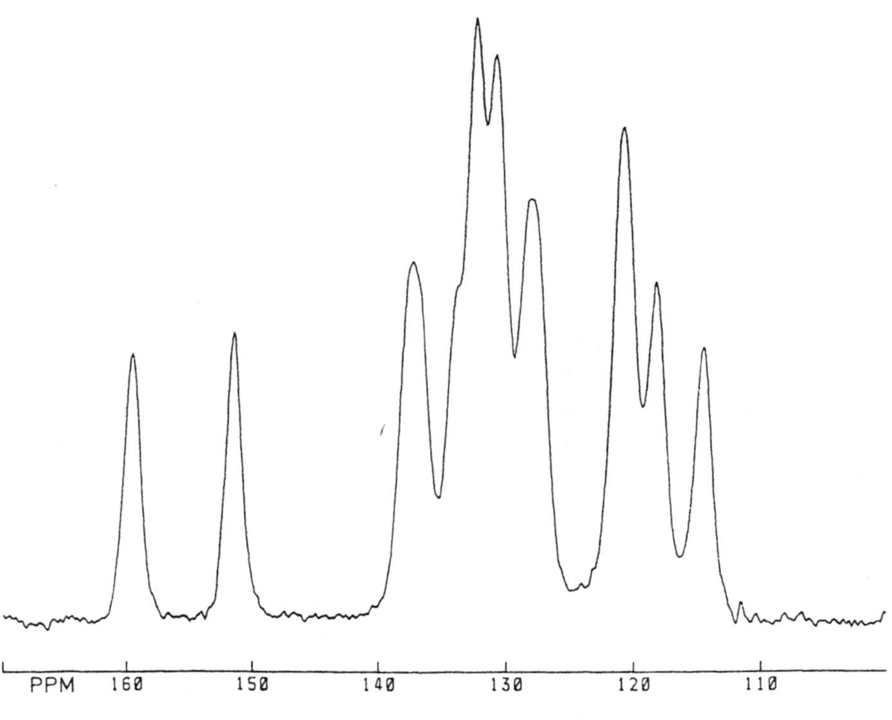

Fig. 5. Carbons *ortho* to ether substituents in an aromatic ring often show inequivalence in the solid state as demonstrated by the ^{13}C spectrum of the oligomer, ϕ-CO-ϕ-O-ϕ-O-ϕ-CO-ϕ

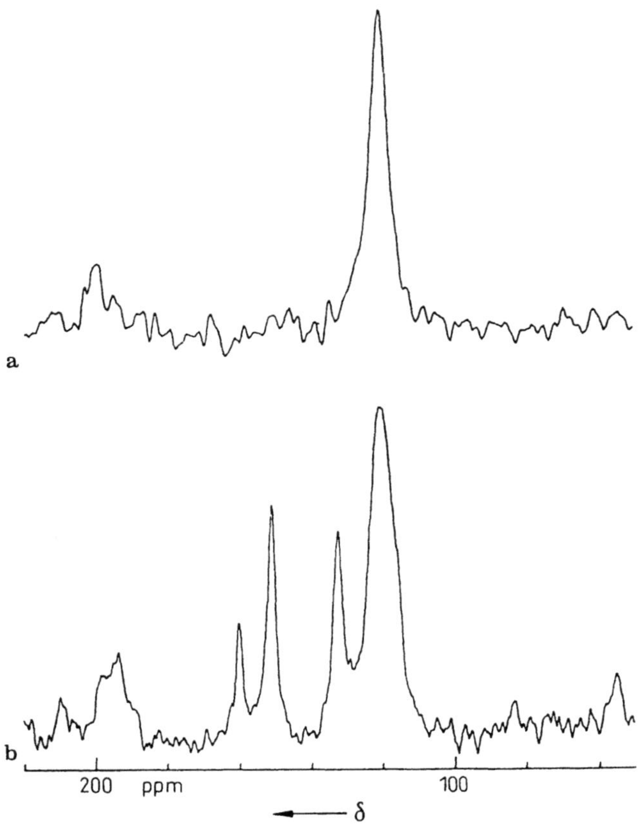

Fig. 6a. Delayed short contact time ^{13}C CP MAS NMR spectrum of PEEK-d_8 indicating the absence of resolved *ortho* carbon inequivalence **b)** Normal ^{13}C CP MAS NMR spectrum of PEEK-d_8, using a contact time of 1 ms

Looking at the resulting spectrum, Fig. 6, we can see no evidence for a splitting of the *ortho* carbon resonance. One problem is that the linewidths of 137 Hz are sufficiently large to prevent the resolution of a small splitting, indeed the *ortho* carbon resonance has over twice the linewidth of the quaternary resonance. Given the *ortho* carbon resonance linewidth we can estimate that the splitting cannot exceed 3.0 ppm thereby placing one limit on the phenyl ring—oxygen torsion angle.

To a large extent conformational analysis has only been applied to the crystalline phases of semi crystalline phases owing to the much greater linewidths in amorphous phases which tend to swamp the small changes in chemical shift. Moreover, in an amorphous phase we can expect so see a greater range of conformational effects so that for example in a polyaromatic there may well be a range in the torsion angles. Only by turning away from the multicentre interactions involved in the chemical shift to two centre dipole–dipole interactions for example can we expect more detailed conformational information. By analogy with solution NMR, conformational analysis is best performed using scalar coupling through expressions like the Karplus equation [37] or dipolar couplings through nuclear Overhauser enhancements [38]. No methods currently exist which allow a small scalar coupling of ca 10 Hz to be resolved in a broad resonance with a linewidth of 300 Hz or more. While for most synthetic polymers as opposed to polypeptides there is no resolution of resonances in similar chemical environments but differrent parts of the polymer chain to allow nOe type experiments.

In conclusion solid state NMR finds little application in industry as a method of conformational analysis. Most work to date has been background studies on semicrystalline polymers correlating NMR spectral features with the known crystal structures. Furthermore, I am only aware of a limited use of the NMR spectrum as a fingerprint of the solid-state packing during polymer treatments.

4 Polymer Orientation

Orientation may be present in a polymer either by design through drawing of a film or by accident as a consequence of the moulding [39]. The most important consequence of orienting a polymer is that the tensile strength will be improved in the direction of the orientation. However, this gain in strength is usually achieved at the expense of poorer mechanical properties perpendicular to the orientation. For a film the orientation induced by either uniaxial or biaxial drawing is advantageous while the orientation in the skin of an injection moulding is generally detrimental. In view of the importance of polymer orientation its determination is essential, especially when trying to understand the effects of processing.

Many techniques are available for measuring the orientation in a polymer such as birefringence, UV light spectroscopy, IR dichroism and X-ray diffraction. Of these birefringence is of the greatest practical use, offering a quick and straightforward measure of the order within a sample. NMR methods offer particular advantages which have not been greatly explored. These are (a) thick samples can be studied (b) the order in crystalline and amorphous phases can be separated and (c) low degrees of ordering can be observed. I will concentrate on these particular strengths.

To study orientation by NMR we must make use of the anisotropic character of the nuclear spin interactions. Early work tended to concentrate on the dipolar type lineshapes of ^{1}H NMR but these are difficult to interpret, involving accurate calculations of the ^{1}H–^{1}H dipolar coupling using a crystal structure [40]. By and large these measurements have been superseded by ones based on the chemical shift anisotropy and quadrupole interaction. These require that the principal components of the tensor be known but not necessarily the crystal structure. Examples of the use of the chemical shift anisotropy are restricted to

^{13}C [41] ^{19}F [42] whilst only ^{2}H [43] has been used for the quadrupole interaction. ^{2}H NMR suffers from the severe disadvantage of requiring isotopic enrichment and does not lend itself to a serious study of processing. For ^{13}C the problem is one of overlapping lineshapes from carbons in different chemical environments, however, this can be overcome in particular cases by using selective experiments. One of the most significant recent developments has been a two dimensional experiment involving MAS which restores the resolution of different carbon sites and in principle widens the applicability of ^{13}C based methods [14]. Again ^{13}C labelling of specific sites could be used without MAS, only the task and expense of producing enough labelled polymer to produce a film on production equipment is daunting. While ^{19}F suffers from severe dipolar broadening like ^{1}H many fluoropolymers, such as PTFE contain ^{19}F in an environment with a large chemical shift anisotropy. If the dipolar broadening is eliminated by multiple pulse sequences the resulting ^{19}F chemical shift anisotropy powder patterns are ideal for determining the orientation [42]. Note though that in general a fluoropolymer has more than one chemical environment and we are back to the problem of overlapping powder patterns seen for ^{13}C. In summary the broadline NMR methods for studying polymer orientation are of restricted value either because isotopically enriched samples are required or because of overlapping lineshapes from different types of groups. However, as I will describe for polyethylene both ^{13}C [44] and ^{2}H[43] NMR techniques are extremely powerful. Again in the appropriate circumstances, such as for PTFE, ^{19}F is very valuable. The ability of NMR to give orientational parameters is not in doubt only the ease with which particular polymers can be studied. It is with these concerns in mind that the MAS experiment recently described assumes great importance since it offers a way to study the bulk of polymers with more than one chemical environment. Experience with this method is not sufficiently extensive to say how easy it will be to use in practice. From the description of the experimental method probably the greatest problem is preparing the sample in a form which is sufficiently large to give a reasonable sensitivity and yet will still spin at 1–3 kHz.

Orientation effects are evident in the broadline NMR spectrum because the frequency of the chemical shift, quadrupolar interaction or dipole–dipole coupling depends on the orientation of the molecules with respect to the applied magnetic field. In an isotropic powder where there is a random distribution of the molecular orientation we get the typical powder pattern shown in Fig. 7a. Should the sample be ordered then certain orientations will be favoured and the intensity distribution of the lineshape perturbed, as for example in Fig. 7b. At the simplest level, without any spectral analysis, the presence of order or changes in the degree of order can be found qualitatively from the NMR lineshape. Quantitative data in the form of the orientational distribution function can be obtained by an analysis of the lineshape. It is common to expand the orientational in terms of Legendre polynomials:

$$P(\beta) = \sum_{l} P_{l00} P_{l}(\cos \beta)$$

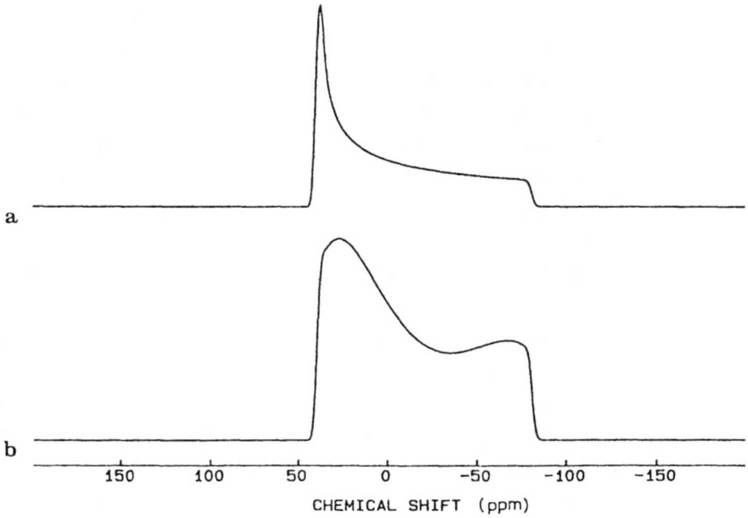

Fig. 7a. Axial chemical shift tensor powder pattern, defined by $\delta = -80$ ppm. **b** Powder pattern of an oriented sample with an axial chemical shift tensor defined by the moments $P_{000} = 1.0$, $P_{200} = 0.6$, $P_{400} = -0.4$, $P_{600} = 0.7$, $P_{800} = -0.5$. The integrated intensity of the isotropic and orientation powder patterns will be equal, in the plot both are normalised to the same height

where P_{l00} are the moments of the orientational distribution and $P_l(\cos\beta)$ describe an increasing degree of order as l increases [45]. Conveniently the NMR lineshape can be calculated as a superposition of subspectra with the subspectra related to the degree of order, see Ref. [43], Fig. 2.

An illustration of the qualitative use of the lineshape is provided by the work of VanderHart on polyethylene [44]. Of particular interest was the use of a selective experiment to observe purely the amorphous region. Changes in this region were followed as a consequence of annealing a sample which had been cold drawn to a nominal draw ratio of 15. The spectrum of the drawn linear polyethylene was consistent with most of the chain segments aligned in the drawing direction. Selection of the amorphous region was based on the shorter carbon T_1^c, typically 2 s, while the crystalline region T_1^c was 700 s. By using the T_1 editing of the spectrum it was possible to show directly that the amorphous phase was ordered; though no attempt was made to quantify this. However, the fraction ordered was substantial, oriented within 15° of the drawing direction. Upon annealing, the oriented material decreased in intensity while the poorer oriented part showed virtually no change in intensity. Two explanations were presented to explain this. First, the annealing only stiffens those chains which have been oriented originally by the drawing process. Secondly, some oriented chains relax during annealing while an equivalent number of poorly oriented chains stiffen or crystallise.

In contrast to the qualitative ^{13}C work a thoroughly quantitative determination of the order in the crystalline parts of a linear polyethylene "single crystal" mat and a drawn sample has been carried out by ^{2}H NMR [43]. An important aim of this work was to show that ^{2}H lineshapes could be fully analysed to give the complete orientational distribution of a partially ordered polymer chain. This was achieved with the orientational distribution found to be essentially gaussian with widths $\beta = 15° \pm 2°$ and $3.4° \pm 0.5°$ for the "single crystal" mat and drawn sample respectively. Subsequent work demonstrated that the order in the amorphous region of the drawn sample was lower. Such work presents a tantalising glimpse at what is possible but broadening of the work into the process area has not been forthcoming because of the need for isotopic enrichment.

Determination of the orientation in a polymer by the MAS technique still relies on the anisotropic character of the nuclear spin interaction only it is manifested differently. We have seen that slow spinning of an isotropic powder results in a centreband at the isotropic chemical shift frequencies of the various resonances and a series of spinning sidebands at integer multiples of the spinning frequency. If on the otherhand we spin an ordered sample, whose order axis is not coincident with the rotor axis, the centreband and spinning sidebands show phase and amplitude modulations depending on the position of the rotor when the sample is pulsed. Hence a sideband pattern in two dimensions can be obtained by synchronising the data acquisition with the rotor position and incrementing the start of the acquisition with respect to the rotor period. An analysis of the intensities in the two dimensional spectrum in terms of the orientation distribution function has been presented and applied to the spectrum of a sample of polyethylene terephthalate fibres [14]. A notable feature of the experiment is that it is sensitive in largely disordered systems and moreover the presence of order is immediately apparent in the spectrum through spinning sidebands in the $\omega 1$ slices. In the particular example studied namely PET fibres the orientational distribution function showed a tilt of about 25° of the ring axis relative to the fibre axis, consistent with earlier X-ray data. Whilst a selective experiment was not carried out to look at the amorphous phase it was concluded that the average ring axis was aligned with the chain axis and where the two deviate the chain axis was still confined to the ring plane.

Despite the obvious suitability of NMR for studying polymer orientation there have been few papers aimed at the more general use of the technique. This cannot be solely attributed to a reluctance of industrial companies to allow details to be published. Rather it points to the lack of emphasis on questions which NMR is strongest at answering. On the whole for the systems being studied birefringence and X-ray diffraction are perfectly adequate and in comparison NMR is somewhat unwieldy. Only when problems in processing require, for example, an accurate, independent measurement of the orientation in an amorphous phase rather than a difference or none at all will NMR become more widely used.

5 Morphology

Solid state NMR can be used to investigate various aspects of polymer morphology such as phase differentiation, identification, selection and domain size. Heterogeneity in a polymer sample is revealed by multicomponent lineshapes or relaxation times. Since the lineshapes and relaxation times vary with temperature because of changes in the polymer motion an accidental equivalence at one temperature can usually be removed by moving to another one. A basic example of phase differentiation is to be able to distinguish between crystalline and amorphous regions in a semicrystalline polymer. We need to understand that the definitions of these regions are not synonymous with X-ray diffraction. In the NMR experiment, take PEK for example, a two component $T_{1\rho}$ is seen with relaxation times of 3.2 ms and 16.5 ms in the ratio of 43:57. Because the crystalline region is rigid, lacking kHz motions, $T_{1\rho}$ is long and we identify this relaxation component with the crystalline phase. However, there is no reason to suppose that this is coincident with the crystalline phase found by X-ray diffraction, by and large it will be, but we can envisage regions in the amorphous phase where tight packing restricts the molecular motion to give a long $T_{1\rho}$. In essence phase differentiation is based indirectly on the molecular structure through the polymer dynamics in contrast to the direct link of X-ray diffraction.

A major advantage of the NMR view of phases is that heterogeneity in a fully amorphous material can be observed and then by use of an appropriate selective experiment the different phases can be identified. The simple view of each phase having a characteristic set of relaxation times is valid only for nuclei of low natural abundance or nuclei giving weak dipole–dipole coupling. When the spin is abundant and has strong dipole–dipole coupling we must take into account spin diffusion [46]. This diffusion process involves the transfer of magnetisation through energy conserving mutual spin flip–flops and leads to the averaging of the relaxation times. Essentially if the spin diffusion can transfer magnetisation between regions with different relaxation rates in a time of the order of the relaxation time a common relaxation time will be seen. Rather than being an impediment to the study of polymer morphology spin diffusion greatly widens its scope by providing a method of looking at domain sizes [47]. Each of the relaxation times, T_1, $T_{1\rho}$, T_2 and T_{1D}, are sensitive to different domain sizes. Assuming a value of the diffusion coefficient we can place lower limits on the size of a domain if separate relaxation times are to be seen. This qualitative view provides a sensitive test of microcompatibility in polymer blends [48]. A quantitative determination of domain sizes is possible by establishing a magnetisation gradient and monitoring the spin diffusion across this gradient.

5.1 Data Analysis

A proper statistical treatment of the data is critical in the application of NMR to polymer morphology. Take as an example the compatibility of two polymers in a blend, a single exponential relaxation decay curve would show compatibility while a double exponential would indicate incompatibility on the characteristic scale of that relaxation time. Further refinements are possible, what if a three exponential fit is statistically significant, then a model involving an interfacial region becomes attractive. In fact mathematical modelling of the spin diffusion process suggests a series expansion of exponentials with variable amplitudes and time constants just for a two phase structure [49]. Only by using measures such as the χ^2 statistic and the goodness-of-fit test can we be confident of not over interpreting the data [50]. From the final points on the plateau of a saturation recovery curve we can determine the standard deviation of the experimental error allowing the resulting χ^2 to be tested. For any given experimental standard deviation we can expect the best fit to correspond to $\chi^2 \approx N$ where N is the number of data points. If $\chi^2 \ll N$ then we are over fitting the data whereas if $\chi^2 \gg N$ the model chosen to fit the data is wrong. The statistical test of the χ^2 is the goodness-of-fit, it is not correct to blithely assume that the lower χ^2 the better. In the goodness of fit we find the probability that an experimental χ^2 as bad as that seen is to be expected given the experimental error and the correct model. Once the goodness-of-fit is unity a decrease in χ^2 means little. We can only attempt a further analysis by improving the experimental data and in so doing reduce the experimental error. Fitting of the data points to a multiexponential process is straightforward using a non linear least squares fitting program. An illustration of the analysis required is the $T_{1\rho}$ behaviour of an amorphous glycerol based polymer shown in Fig. 8. Table 3 lists the results of a one, two and three exponential fit. For the single exponential, χ^2 is very large and Q is

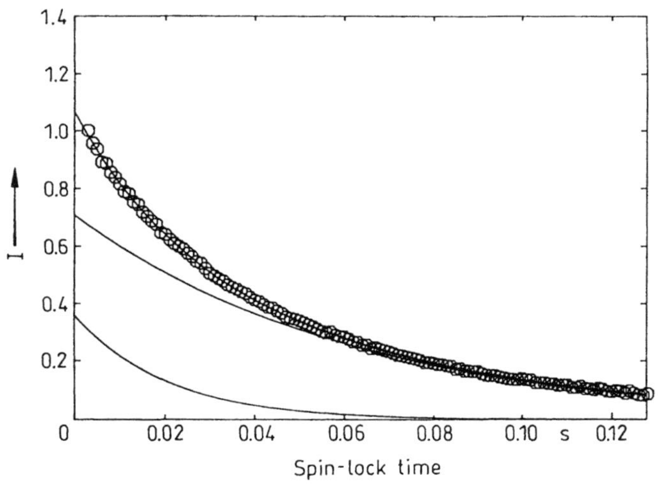

Fig. 8. $T_{1\rho}^{H}$ decay acquired by a single point method, together with a two exponential fit obtained using a non-linear least squares fitting program. The best fit corresponds to $I_1 = 0.34$ $T_{1\rho}(1) = 19$ ms $I_2 = 0.66$ $T_{1\rho}(2) = 60$ ms

Table 3. Results of a non linear least squares fitting to the $T_{1\rho}$ decay of an amorphous polymer

Number of exponentials	1		2		3	
	Fraction	$T_{1\rho}$(ms)	Fraction	$T_{1\rho}$(ms)	Fraction	$T_{1\rho}$(ms)
					0.77	40.6
	1.0	47.5	0.66	59.8	0.11	8.9
			0.34	20.0	0.14	123.5
χ^2		2065.0		127.7		112.2
Q		0.0		0.39		0.76

zero, clearly the model is wrong. However, both the two exponential and three exponential fits are better with acceptable χ^2 as indicated by the reasonable Q. Which should we prefer the three exponential because it has the lowest χ^2 or the two exponential because the reduction in χ^2 is not statistically significant in which case we should apply Occams razor and choose the simpler hypothesis? One way to test the significance of the difference in χ^2, $\Delta\chi^2$, is to use the same incomplete Gamma function as for Q which describes the distribution of χ^2. In this case the χ^2 seen for the three exponential fit lies within one standard deviation of the expected value and hence does not provide support for rejecting the two exponential model. The reduction in χ^2 for the three exponential model has no statistical significance.

We can at this stage include the results from other techniques which might favour the three exponential model but we must be clear that NMR is now not providing independent evidence of this; we are choosing a more sophisticated model on the basis of other measurements.

An important element of considering the statistical significance of the number of exponentials used to represent a relaxation profile is the distinction between different relaxation times. Thus if a $T_{1\rho}$ decay consists of a two component decay of equal amplitudes with values of 20 ms and 40 ms it might in fact be a single component decay with a value of 30 ms. The problems then are; first the number of exponentials necessary to model the data and secondly when the errors of the fitted parameters are taken into account do the two time constants overlap. By modelling the fitting process with generated data it can be shown that even with noisy data, a 5% experimental standard deviation, in the above example the time constants with their 95% confidence limits do not overlap, $T^1_{1\rho} = 18.3 \pm 3.3$ ms and $T^2_{1\rho} = 38.4 \pm 3.3$ ms. On the otherhand the difference in χ^2 between the single and double exponential models only becomes statistically significant when the standard deviation of the experimental intensity is less than 1%. As the proportion of the two components changes, with one becoming a minor component, the signal-to-noise and spectrometer stability required will also increase. For a 20% component the experimental error must be ca 0.5% to be able to be confident about its presence if its relaxation time is only twice the dominant component.

5.2 Crystallinity

One of the simplest measurements which we can hope to perform is to determine the crystallinity of a simple. Various techniques have been used to measure the crystallinity such as, wide line X-ray diffraction, density, heat of fusion, NMR and laser Raman. Given the variety of techniques, based on differing physical models of what characterises a crystal phase it is not surprising that inconsistent results can be obtained. It has been argued though that such discrepancies can give useful information [10]. NMR methods all rely on being able to uniquely identify one particular relaxation time with a crystalline phase. This may be the carbon T_1 as illustrated by work on polyethylene [51] the proton $T_{1\rho}$ as in the case of PEEK, or the proton T_{1xz} for PET [52]. Having identified the unique relaxation parameter, the measurement of the crystallinity is then just a matter of measuring and fitting the relaxation profile to extract the intensity of the desired component. Complications arise when the phase model is in doubt, raising questions such as, is there an interfacial region and if so what are its relaxation parameters.

If a simple experimental procedure including data analysis is possible for the polymer in question then NMR can be used to provide a routine determination of the crystallinity. Moreover, if it is based on wideline ^{1}H, low field systems such as the Bruker Minispec can be used and automation of the analysis is feasible. Indeed it is this possibility of a routine test measurement which makes NMR an attractive method for determining the crystallinity of a polymer in a manufacturing process.

5.3 Domain Size

Qualitative domain sizes can be estimated assuming a spin diffusion co-efficient, D_s, and Eq. (1) [53]. An approximate value for Ds in a rigid hydrocarbon is 10^{-15}–$10^{-16}\,m^2s^{-1}$ [49]. Hence if the ^{1}H T_1 is of the order of 1 s and only a single component seen any domains must be smaller than 200 Å conversely if a multicomponent T_1 is observed the domains must be larger than this. Since $T_{1\rho}$ are shorter than T_1 their equivalence or otherwise is sensitive to a smaller domain size

$$\langle r^2 \rangle \sim (6D_s\tau)^{1/2} \quad (1)$$

Modifying the relaxation time through changing the temperature, altering the static field B_o or the effective B_1 field in a $T_{1\rho}$ experiment offers the possibility of widening the qualitative limits on the domain size.

More exact domain sizes can be found by observing the transfer of magnetisation along a gradient. The key feature is establishing the magnetisation gradient in which the magnetisation in one particular type of domain is selectively destroyed. How this is done depends on whether T_1, $T_{1\rho}$ or T_2 are different between the phases. One of the most popular methods has been the Goldman–Shen pulse sequence which is applicable when differences in T_2 exist [53]. The recovery in the magnetisation through spin diffusion has been modelled for polyethylene and polypropylene assuming the diffusion to be 1-D, 2-D or 3-D [47]. However owing to the small differences in the plots of recovered intensity as a function of the delay for the different cases distinguishing between them was difficult. A further complication is that diffusion might not be isotopic. Although the domain size is critically dependent on the model chosen for the diffusion e.g. for unconditioned polypropylene film 1D, 8 Å; 2D, 20 Å and 3D 33 Å a consistent picture could be seen comparing the unconditioned and conditioned samples because for each dimensional model of diffusion the conditioned sample gave a larger domain size. This was noted as being consistent with previous observations concerning the increase in the non crystalline domain upon annealing.

Fewer experiments have been reported exploiting differences in $T_{1\rho}$. One reason is that the straightforward analogous experiment to the Goldman–Shen is complicated by spin diffusion which still occurs even during the spin-lock. Using a preparation period of 1 ms would lead to spin diffusion over a distance of 10 Å blurring any domain size information obtained. To overcome this a multiple pulse method has been proposed and applied to PET fibres to study morphology on the 10–50 Å scale [52]. From a three exponential fit to the multiple-pulse relaxation time T_{1xz} the authors conclude that the two phase crystalline-amorphous model is a poor representation of the PET fibre morphology: yet no data is presented to show the statistical basis on which the three exponential model is to be preferred over a two exponential one. One is not in a position to judge whether the NMR data have been over interpreted to conform with X-ray results or whether here we have clear evidence to support a more complex morphology. In view of these uncertainties the subsequent discussion of domain sizes becomes weakened.

Experiments to date demonstrate that solid state NMR can provide information on the domain sizes in polymers. More importantly the bulk nature of the technique complements microscopy where domains in only small regions of a sample can be imaged. Despite this there does not appear to be a widespread use of NMR to determine domain sizes in general.

5.4 Phase Identification

After demonstrating that differentiation of phases in a heterogeneous sample is possible based on differences in the relaxation characteristics it is a logical step to wish to define the composition of the phases. There is value in this even for semicrystalline polymers where the distinction is between a crystalline and an amorphous phase. Take as an example an ester copolymer based on hydroxybenzoic acid. Are the compositions of the crystalline and amorphous phases similar or is there a tendency for the copolymer to become blocky with the blocks in one phase or another? Such crystalline, amorphous phase selection is usually quite simple because of differences in the $T^{H}_{1\rho}$ for the two phases. Thus the phase with the longer $T^{H}_{1\rho}$, normally the crystalline phase, can be selected by using a delayed contact time experiment. It is not so straightforward to select the shorter $T^{H}_{1\rho}$ phase. We have seen earlier though that it is still possible to select an amorphous phase based on differences in T^{c}_{1}.

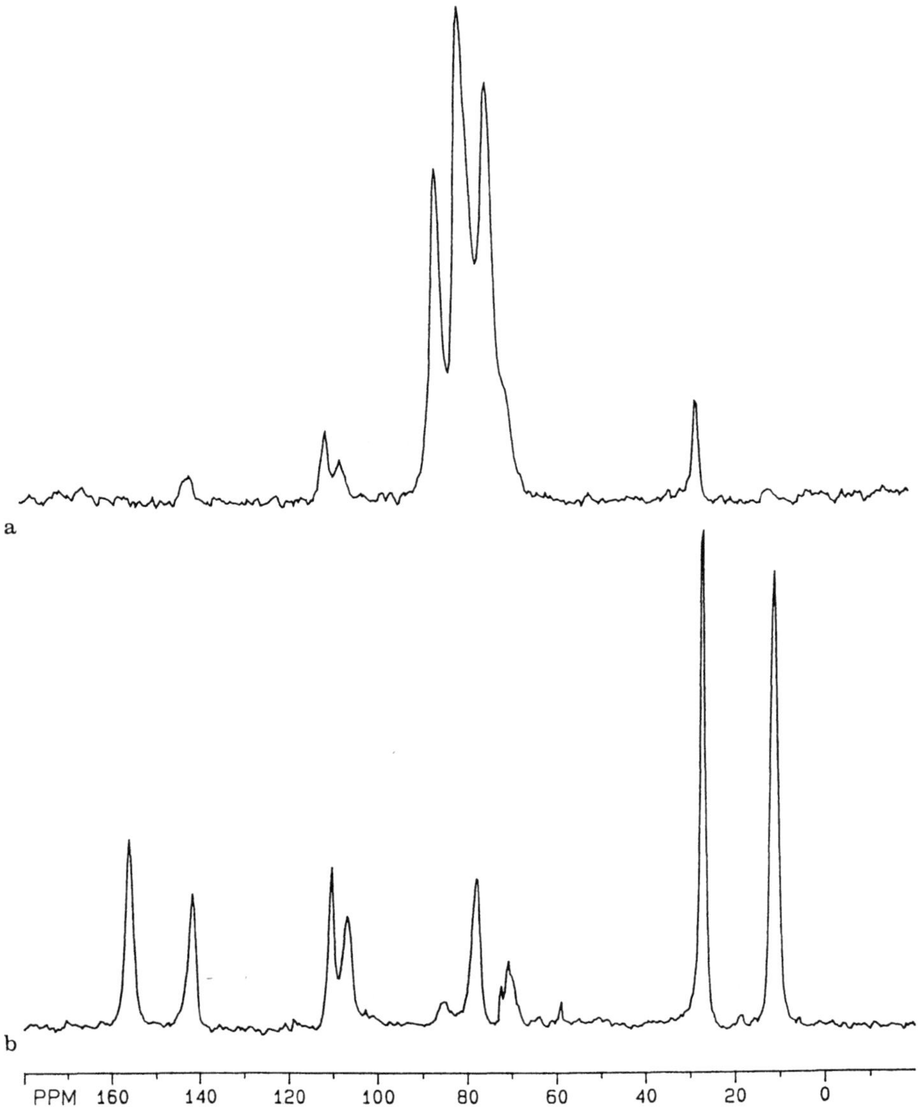

Fig. 9 a, b. Phase selection in a heterogeneous amorphous polymer based on the differences in mobility in the two phases. **a**) A short contact ^{13}C CP MAS NMR spectrum highlighting the rigid matrix **b**) Dipolar dephased ^{43}C CP MAS NMR spectrum showing the more mobile component

A potential area for phase identification is that of polymer blends where we can consider determining the composition of different phases and show the extent of the solubility of one polymer in another. The main restriction on the accuracy of any quantitative results will be the ratio of the relaxation times for the phases. When the ratio approaches unity no selection of the phases will be possible but as the ratio increases so the phase selection will improve as will the signal-to-noise possible in the selective experiment. The ability to resolve different phases and identify them perhaps has the greatest utility when looking at an amorphous blend or other heterogeneous amorphous material. Not only the bulk polymer can be studied but also where a compatibiliser ends up if the blend remains incompatible on the local level. A good example of the phase selection possible in a heterogeneous amorphous material is provided by a polymer formed by the reaction of glycerol, and furfuryl alcohol with a metal alkoxide. The ^1H NMR clearly shows a two component free induction decay corresponding to a short and long T_2 component. In the long T_2 component molecular motion greatly reduces the carbon-hydrogen dipolar coupling decreasing the efficiency of cross-polarisation. It is therefore possible to select the rigid matrix corresponding to the short ^1H T_2 by using a short contact time in the cross-polarisation pulse sequence, Fig. 9a. On the other hand to select the mobile component we can make use of an interrupted decoupling or dipolar diphased spectrum, Fig. 9b. Clearly the rigid matrix is a cross-linked glycerol one while the mobile phase is a furfuryl alcohol, metal alkoxide based one.

6 Polymer Dynamics

The aim in studying the dynamics of a polymer chain is to provide a repeat-unit level characterisation of the local motion which can be related to the physical properties of the polymer. Much of the emphasis in the literature is on the first stage, properly determining the types of motion and their frequency and amplitude, with the consequence that the more important property relationship has been neglected. From the standpoint of the use of solid state NMR in the polymer industry this is not really acceptable. There must be a firm relationship between the polymer dynamics measured by NMR and the properties of the polymer. In a series of polymers containing phenylene rings a strong correlation has been established between the presence of ring flips and the impact strength of the polymers [54]. But there is debate about this link because a pure π flip would not be mechanically active hence the ring motion is only monitoring the active dynamic process. Understanding the relationship between the polymer motions and the mechanical properties is clearly facilitated by working on well established products such as polycarbonate but it is equally apparent that these studies are not going to markedly affect the business. In view of the more fundamental nature of work on polymer dynamics the methods must be applied early in the development of new materials. One possibility is to screen new polymers at the laboratory stage.

Building upon the idea that only polymers with phenylene ring motion show a high impact strength regardless of the detailed link we can study aromatic polymers composed of various linkages and establish which combinations are good. This is not as straightforward as it appears because an ether–ether ring will flip in the amorphous phase of PEEK (poly ether–ether ketone) but not in poly phenylene oxide. Both intermolecular and intramolecular interactions need to be taken into account in polymer design. We can view the polymer dynamics as reporting on the packing in the solid when the intermolecular terms are important. By using the structural specificity of NMR we can say which groups contribute to the dynamic processes as a function of temperature and say whether the intermolecular interactions have altered the initial considerations.

At present empirical rules are used to predict the effect of various groups on the T_g of a polymer and the dynamics [55]. For example a biphenyl unit is rod-like and will raise the T_g but does it possess the ring dynamics to prevent the polymer becoming brittle? Then what is the effect of a *meta* link? Is the *para* ether–ether ring constrained in the polymer?

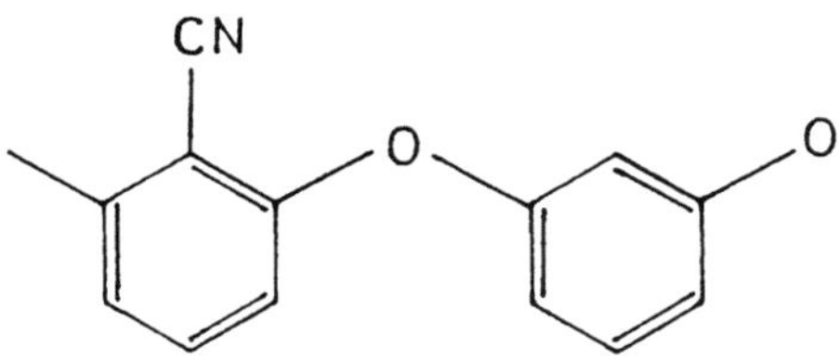

T_g = 147 °C

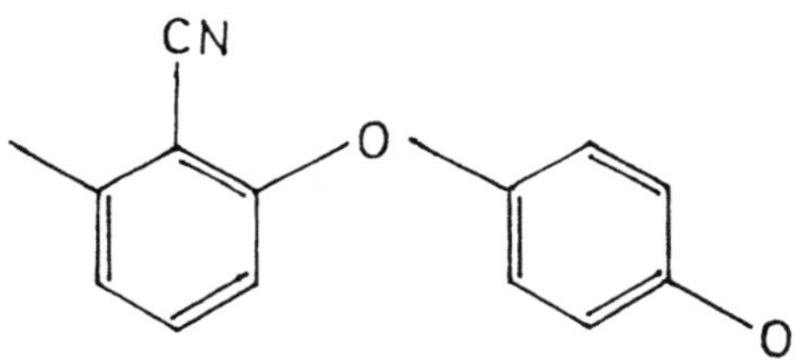

T_g = 174 °C

With existing systems the effect of additives could be studied but when compared with the direct mechanical testing it is obvious that NMR must be aimed at finding the underlying molecular level mechanism.

In terms of the depth of information which can be obtained NMR is unique and is simply the best technique for studying polymer dynamics. When considering the relationship to the mechanical properties NMR complements dynamic mechanical analysis by providing a molecular level interpretation of the loss peaks. Two types of NMR measurements can be used to examine the polymer dynamics; nuclear spin relaxation times and lineshape analysis.

6.1 Nuclear Spin Relaxation Times

Relaxation time measurements are usually carried out on the ^{1}H spins but other nuclei present can be used. The principal advantage of proton NMR is the sensitivity which allows a high signal-to-noise decay curve to be acquired enabling a more accurate determination of the relaxation time, particularly if the decay is multicomponent. ^{13}C relaxation time measurements are commonly performed with cross-polarisation and magic angle spinning giving site specific data [56]. Four relaxation times are often measured; T_1 the spin-lattice relaxation time, $T_{1\rho}$ the spin-lattice relaxation time in the rotating frame T_2 the spin–spin relaxation time and T_{1D} the dipolar order relaxation time. These vary with the temperature as illustrated in Fig. 10. Each of the relaxation times is sensitive to motion at different frequencies and thus a wide range of motional frequencies can be studied over the accessible temperature range. Given the comment made earlier regarding the importance of kHz frequencies on the mechanical properties corresponding to $T_{1\rho}$ one might question the point of measuring all the relaxation times. As a minimum T_1 must also be measured to ensure that the $T_{1\rho}$ experiment is properly set up, especially if the sample is heterogeneous. Since T_1 is sensitive to MHz frequencies it is little value for the mechanical properties. On the other hand T_{1D} is sensitive to slow motions and in the strong collision limit is related to the jump frequency Θ_D by $T_{1D}^{-1} = k\Theta_D$ [57]. Thus T_{1D} can indicate kHz motions which are not accessible by $T_{1\rho}$ within the temperature range of the probe. Furthermore slow motions should be

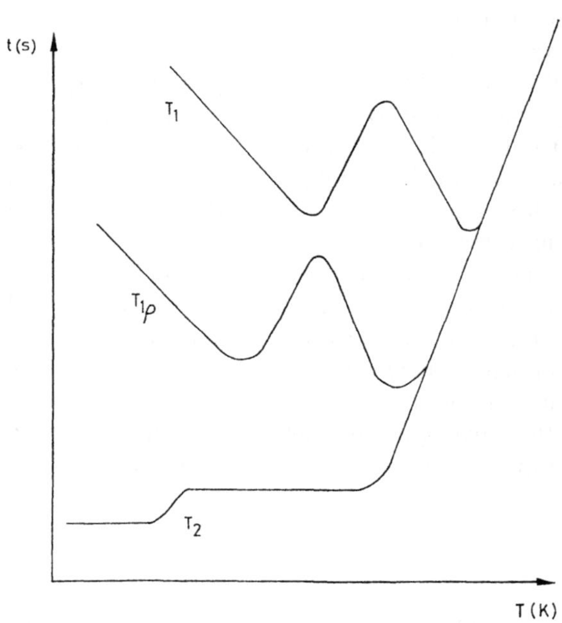

Fig. 10. Schematic variation in nuclear spin relaxation times as a function of temperature

relevant to the long term response of a polymer under load. The least justification can be found for T_2 but given the short time required to measure it, only a few minutes, it warrants inclusion, giving as it does evidence of large amplitude motions. The characteristic features of the variation of relaxation times with temperature shown in Fig. 10 are; (1) T_1 and $T_{1\rho}$ show minima at which temperature the frequency of the motion is given by $\sim\omega_0$ and $\sim N(\gamma B_1)^{-1}$ respectively. (2) A transition in T_2 which occurs when molecular motion averages the dipolar coupling corresponding to a motional frequency of 10–100 kHz. (3) More than one T_1 or $T_{1\rho}$ minimum can be seen and more than one T_2 transition. Each represents a different type of motion. When a number of motions are possible the various minima or transitions can overlap greatly complicating the interpretation of the data. Owing to problems with spin diffusion, it is difficult to interpret the T_1 or $T_{1\rho}$ minima in terms of the motion of specific groups. Rather than observing separately a T_1 or $T_{1\rho}$ for each group only an average relaxation time is seen negating the structural specificity of NMR [58]. This can present an incentive to study the ^{13}C relaxation times since these do not suffer from averaging by spin diffusion. Providing the relaxation time temperature dependence can be identified with a single type of motion the activation energy can be obtained by determining the frequency of the motion from analytical expressions for the relaxation times. Most work has attempted to correlate the dynamical information found by NMR relaxation time analysis to dielectric and dynamic mechanical relaxation. Generally speaking the correlation is excellent but it must be noted that if the group specificity is lost in the NMR measurements because of spin diffusion so to is its special advantage and the identification of different frequencies with particular types of motion is entirely indirect.

Schaefer and coworkers have made a case for the use of $T^{c}_{1\rho}$ to characterise the presumed to be mechanically active kHz main chain motions of amorphous polymers [59], their case being strengthened by the difficulty with ^{1}H measurements. Unfortunately $T^{c}_{1\rho}$ can reflect spin–spin interactions arising from the carbon spins coupling to protons when the $^{1}H-^{1}H$ dipole coupling is strong and unmodulated by motion. This relaxation pathway is unrelated to the main chain motions. In view of this Schaefer presented a protocol, involving the rise time constant T_{CH} in the ADRF cross polarisation experiment, to ensure that the $T^{C}_{1\rho}$ is determined solely by motional processes. Earlier work established a good correlation between the ratio $\langle T_{CH^+}\rangle/T^{c}_{1\rho}$ for protonated main chain carbons and the Izod notched impact strength, where $\langle T_{CH^+}\rangle$ is the average spin-lock cross polarisation rise time constant corrected for the relative $^{1}H-^{1}H$ dipolar interaction. Despite this work there has been little follow up, such as studying the variation in this ratio with sample preparation or the presence of additives.

6.2 Lineshape Analysis

Relaxation time measurements are able to give the frequency and activation energies for a dynamic process but they do not tell us the type of motion nor its amplitude. For this we need to turn to lineshape analysis, using the broad powder patterns introduced earlier. As a result of molecular motion characteristic changes take place in the lineshape. Only the chemical shift anisotropy and quadrupole powder patterns are of value. This is because they are intrinsic and do not depend on intermolecular interactions as does the dipolar coupling. Being intrinsic they report uniquely on the reorientation of that particular nuclear spin. The most commonly used lineshapes are the ^{13}C chemical shift anisotropy and the ^{2}H quadrupolar ones. Both have disadvantages; in a carbon based polymer it may be difficult if not impossible to obtain the powder pattern of just one carbon type, though selective experiments can help here. In some cases ^{13}C isotopic enrichment may be the only method. With ^{2}H isotopic enrichment is essential given the low natural abundance. Although I have presented the requirement for isotopic enrichment as a disadvantage, providing the will to enrich is present, a test perhaps of the business commitment to the problem, this can be an immense advantage because by selective enrichment tagging of only those groups of interest is possible.

For most purposes the ^{2}H NMR methods are more versatile than the ^{13}C ones. One exception to this is when the axial character of the ^{2}H quadrupole interaction prevents two motional models being distinguished then the non-axial character of the chemical shift anisotropy of most carbon environments could prove useful. The two particular advantages of ^{2}H NMR are the sensitivity to a greater range of motional frequencies, 1 Hz–10 MHz, and the greater ease of synthesising isotopically enriched samples. Rather than discuss the breadth of lineshape analysis I will present results on one polymer as an example to illustrate the methodology. The wider view is available in a number of reviews and books. [10, 60].

To acquire a ^{2}H NMR spectrum which is as undistorted as possible an echo sequence must be used to overcome the problem of the receiver dead time. This basic echo sequence is usually combined with a saturating train of pulses and a delay for the magnetisation recovery providing both a phase selection and a frequency selection method. By applying a saturating train of pulses all the net magnetisation is destroyed while during the delay the magnetisation recovers according to the ^{2}H T_1. As we have seen above the T_1 reflects the frequency of

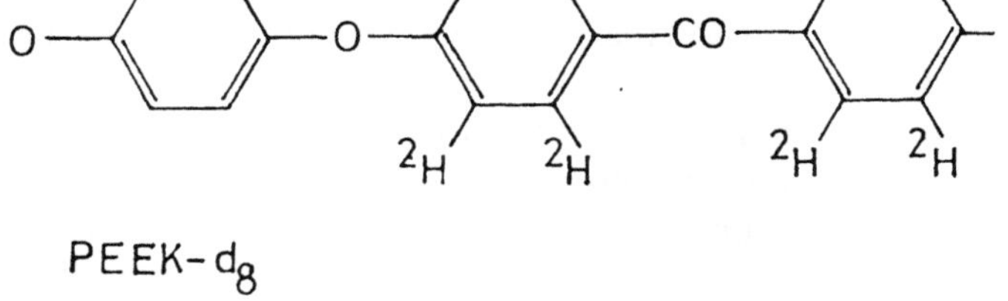

PEEK-d_8

the motion, decreasing with frequency until the T_1 minimum is reached. Thus in an amorphous polymer where we can expect a distribution of frequencies so we can expect a distribution in T_1, moreover, in contrast to ^1H NMR spin diffusion does not average the T_1 values. Hence if we leave a short recovery time the ^2H spectrum will be biased towards the faster moving sections of the polymer. At one extreme, in a semicrystalline polymer, the crystalline phase may be sensibly rigid with a long T_1, it is then generally going to be possible to set the delay period to get a representative spectrum of the amorphous region which will have a shorter T_1. Note that even here because the measured T_1 for the amorphous phase will be an average there will be a tendency to bias the spectrum towards a faster motion. This is clearly the case for an amorphous polymer if we leave a recovery delay of much less than the average T_1. Thus the first step must be to measure the ^2H T_1.

As expected for a semicrystalline polymer a two component T_1 is seen for PEEK-d_8, a shorter T_1 from the amorphous phase of 18 ms, and a longer one of 1.6 s for the crystalline phase. This must be repeated at each temperature the polymer is studied. To select the amorphous phase a recovery time of $3T_{1a}$ is used, such a delay leads to a negligible contribution from the crystalline phase.

At the two extremes of motion we have the static lineshape, Fig. 11a and the fast limit lineshape, Fig. 11c. Depending on the nature of the motion different fast limit spectra will be seen, the spectrum shown is for a phenylene ring flip. At intermediate rates of motion a more complex lineshape is seen arising from the effects of the echo pulse sequence [61].

Before comparing these types of lineshapes with the experimentally observed lineshape, Fig. 12, two additional factors must be taken into account. First, corrections must be made to the calculated spectra to take into account the finite power of the pulses in the echo sequence [62]. Secondly, because of the heterogeneous nature of amorphous polymers a distribution of flipping rates must be used. Two common distributions are a log normal [63] and a William-Watts fractional exponential [64], similar results have been reported for the two. The parameters describing the distribution are found by comparing the observed spectra with calculated ones. Some indication of the breadth of the distribution is given by the reduction in the echo amplitude. As the distribution becomes broader so the variation in the echo amplitude with temperature becomes flatter. Having obtained a fit between the observed and calculated spectra for one echo delay time this should be confirmed by using a different echo delay.

Only after carrying out this procedure for a range of temperatures can an activation energy be determined. But for each temperature we will know the mean frequency, the breadth of the frequency distribution and the type of motion taking place. To complicate matters further more than one motion might be present, indeed it is common to find torsional oscillations occurring simultaneously with ring flips [60]. However, as the work of Spiess shows these can be taken into account. It cannot be denied that a thorough analysis of molecular dynamics by ^2H NMR is time consuming both from the standpoint of

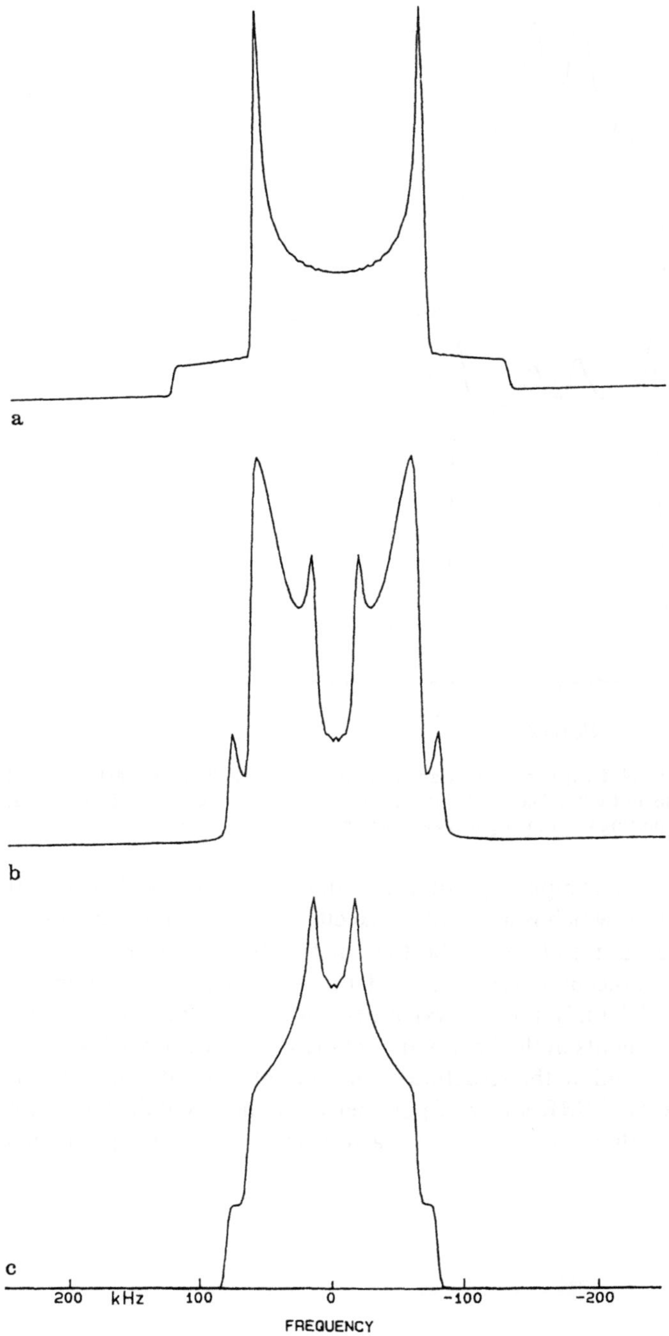

Fig. 11 a–c. ^2H quadrupole echo lineshapes for different motional cases depending on the echo delay time τ_e and the correlation time for the motion. τ_c The motion is a phenylene ring π flip. **a**) slow limit, $\tau_c \gg \tau_e$ **b**) intermediate, $\tau_c \sim \tau_e$ and **c**) fast limit $\tau_c \ll \tau_e$

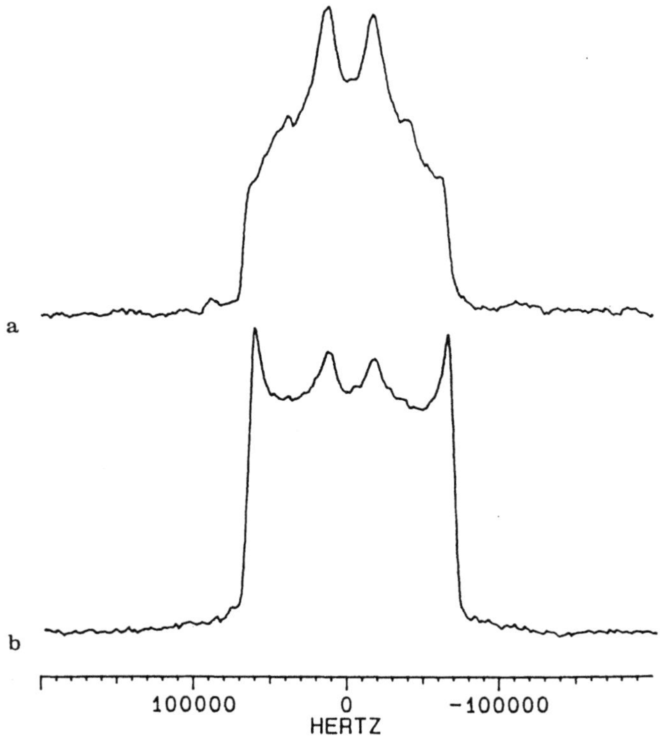

Fig. 12. ^2H quadrupole echo NMR spectrum of the amorphous phase of PEEK-d_8 at 300 K acquired using an echo time of 30 μs **b)** Fully relaxed ^2H quadrupole echo NMR spectrum of PEEK-d_8 at 300 K showing the outer splitting from the rigid crystalline phase

spectrometer time and the computing required. Equally it cannot be denied that the dynamic information which is available is excellent, unrivalled by any other technique. A realistic appraisal would be that in the absence of a clear link between the detailed molecular dynamics and the properties of a polymer a thorough going ^2H NMR analysis is not warranted within a commercial concern. However, with developments in the field of modelling polymer properties NMR will be needed to help validate the structural models used. Although my view is that a comprehensive ^2H NMR study of polymer dynamics by industry is not appropriate at present there are cases where a restricted range of experiments can be useful.

7 Conclusion

Solid state NMR is now established as an invaluable technique for the characterisation of insoluble polymers, providing detailed information on the types of structural groups present, extent of curing and the nature of cross-linking sites. Although proving experiments have been carried out in a number of areas, notably in the study of orientation and polymer morphology, NMR has not made significant inroads into the more traditional methods of examination. Indeed NMR can only be expected to be usefully applied when its particular strengths are relevant such as in determining the orientation of individual layers in a laminated film where the chemical selectivity of the NMR spectrum can be used to good effect. In essence the strengths of NMR are the ability to identify polymer structures based on their high resolution NMR spectra and then to select these constituents in a multi-phasic material on the basis of the chemical shifts in the NMR spectrum or on their nuclear spin relaxation times.

As the importance of more complex polymer structures and polymer blends increases so NMR will be more widely applied to study orientation and morphology. The use of NMR to determine the polymer dynamics is of undoubted value, however, it must be stressed that correlations with basic physical properties have not been numerous. Until a firmer predictive link can be made, scepticism will remain about the value of such measurements in the polymer industry.

8 References

1. VanderHart DL, Earl WL, Garroway AN (1981) J. Magn. Reson., 44: 361
2. Dec SF, Bronniman CE, Wind RA, Maciel GE (1989) J. Magn. Reson., 82: 454
3. Mehring M (1983) Principles of high resolution NMR in solids. Springer, Berlin Heidelberg New York
4. Gerstein BC, Dybowski CR (1985) Transient techniques in NMR of solids. An introduction to theory and practise. Academic Orlando, FL
5. Haeberlen U (1976) High resolution NMR in Solids: Selective averaging. New York: Academic
6. Fyfe CA (1983) Solid state NMR for Chemists. Guelph: CFC Press
7. Clayden NJ (1988) Chemica Scripta, 28: 211
8. Schaefer J, Stejskal EO (1979) In: Levy GC (ed) Topics in carbon-13 nuclear magnetic resonance spectroscopy. Wiley, New York, vol 3 p 283
9. Fedotov VD, Schneider H (1989) Structure and dynamics of bulk polymers by NMR methods. Springer, Berlin Heidelberg New York
10. Komoroski RA (ed) (1986) High resolution NMR spectroscopy of Synthetic Polymers in Bulk. VCH, New York
11. Wind RA, Dec SF, Lock H, Maciel GE (1988) J. Magn. Reson., 79: 136
12. Dixon WT (1982) J. Chem. Phys., 77: 1800
13. a) Maricq MM, Waugh JS (1979) J. Chem. Phys., 70: 3300
    b) Herzfeld J, Berger AE (1980) J. Chem. Phys., 73: 6021
14. Harbison GS, Vogt V-D, Spiess HW (1987) J. Chem. Phys., 86: 1206
15. Alemany LB, Grant DM, Pugmire RJ, Zilm KW (1983) J. Am. Chem. Soc., 105: 2133
16. Harris RK (1985) Analyst 110: 649
17. Allen PEM, Simon GP, Williams DRG, Williams EH (1989) Macromolecules 22: 809
18 Fyfe CA, Rudin A, Tchir WJ (1980) Macromolecules 13: 1322
19. Harris RK (1978) NMR and the periodic table. Harris RK, Mann BE (eds). Academic, London
20. Nishiyama N, Horic K, Asakura T (1987) J. Applied Polym. Sci., 34: 1619
21. Clayden NJ, Palasz P (1990) J. Chem. Res., 68
22. Sefcik MD, Stejskal EO, McKay RA, Schaefer J (1979) Macromolecules 12: 423
23. Sadtler Standard Carbon-13 NMR, Sadtler Research Labs. Philadelphia, PA, 1977

24. Kaplan S, Jansen F, Machonkin M (1985) Appl. Phys. Lett., 47:750
25. Jarman RH, Ray GJ (1985) J.C.S. Chem. Commun., 1153
26. Axelson DE (1985) Solid state nuclear magnetic resonance of fossil fuels. Multiscience. Canada
27. Beamson G, Brennan, Clayden NJ, Jennings RCK J. Polym. Sci. Polym. Phys. Ed., (in press)
28. Kaplan S, Dilks A (1983) J. Polym. Sci., Polym. Chem. Ed., 21:1819
29 Munowitz M, Aue WP, Friffin RG (1982) J. Chem. Phys., 77:1686
30. Yannoni CS, Clarke TC (1983) Phys. Rev. Lett., 51:1191
31. Gomez MA, Tanaka H, Tonelli AE (1987) Polymer 28:2227
32. Maciel GE (1974) Topics in Carbon-13 Nuclear magnetic resonance spectroscopy. Vol. 1 Levy GC (ed). John Wiley. New York
33. Grant DM, Cheney BV (1967) J. Am. Chem. Soc., 89:5315
34. Bunn A, Cudby MEA, Harris RK, Packer KJ, Say BJ (1982) Polymer 23:694
35. Garroway AN, Ritchey WM, Moniz WB (1982) Macromolecules 15:1051
36. Clayden NJ (1989) Magn. Reson. Chem., 27: 692
37. Marshall JL (1982) In: Marchand AP (ed) Methods in stereochemical analysis vol 2. Chemie International, Florida
38. Gronenborn AM, Clore GM (1985) In: Emsley JW, Feeney J, Sutchffe LH (eds) Progress in nuclear magnetic resonance spectroscopy, vol 17, Pergamon, Oxford
39. Ward IM (1975) Structure and properties of oriented polymers. New York: John Wiley 1975
40. Cunningham A, Manuel AJ, Ward IM (1976) Polymer 17:125
41. Hempel G, Schneider H (1982) Pure and Appl. Chem., 54:635
42. Brandolini AJ, Alvey MD, Dybowski CJ (1983) J. Polym. Sci., Polym. Phys. Ed., 21:2511
43. Hentschel R, Sillescu H, Spiess HW (1981) Polymer 22:1516
44. VanderHart DL (1979) Macromolecules 12:1232
45. Hentschel R, Schlitter J, Sillescu H, Spiess HW (1978) J. Chem. Phys., 68:56
46. Abragam A (1961) The Principles of Nuclear Magnetism. Oxford University Press, Oxford
47. Cheung TTP, Gerstein BC (1981) J. Appl. Phys., 52:5517
48. McBrierty VJ, Douglass DC, Kwei TK (1978) Macromolecules 11:1265
49. Kenwright AM, Packer KJ, Say BJ (1986) J. Magn. Reson., 69:426
50. Press WH, Flannery BP, Teulolsky SA, Vetterling WT (1986) Numerical recipes. The Art of Scientific Computing. Cambridge University Press, Cambridge
51. Komoroski RA, Maxfield J, Sakaguchi F, Mandelkern L (1977) Macromolecules 10:550
52. Havens JR, VanderHart DL (1985) Macromolecules 18:1663
53. Goldman M, Shen L (1966) Phys. Rev., 144: 321
54. Schaefer J, Stejskal EO, Buchdahl R (1977) Macromolecules 10: 384
55. Van Krevelen DW (1980) Properties of polymers. Elsevier, Amsterdam
56. Torchia DA (1978) J. Magn. Reson., 30:613
57. Slichter CP, Ailion DC (1965) Phys. Rev., 135:A1099
58. McCall DW (1971) Acc. Chem. Res., 4: 223
59. Schaefer J Stejskal EO, Steger TR, Sefcik MD, McKay RA (1980) Macromolecules 13:1121
60. Spiess HW (1985) Adv. Polym. Sci., 66:24
61. Rice DM, Wittebort RJ, Griffin RG, Meirovitch E, Stimson ER, Meinwald YC, Freed JH, Scheraga HA (1981) J. Am. Chem. Soc., 103: 7707
62. Bloom M, Davis JH, Valic MI (1980) Can. J. Phys., 58:1510
63. Spiess HW (1983) Coll. Polym. Sci., 261:193
64. O'Gara JF, Jones AA, Hung C-C, Inglefield PT (1985) Macromolecules 18:1117

Solution NMR of Synthetic Polymers

A. Bunn

ICI plc, Wilton Materials Research Centre. PO Box No. 90, Wilton, Middlesbrough, Cleveland, TS6 8JE, U.K.

Table of Contents

NMR Basic Principles and Progress, Vol. 29

Solution state NMR has for many years been the primary technique for determining the molecular structure of synthetic polymers. This article illustrates how solution state NMR data from modern spectrometers is used for polymer evaluation in an Industrial Laboratory. ^{1}H- and ^{13}C-NMR constitute the majority of the work described and some ^{19}F-NMR data are presented. The author reviews a wide range of polymers and how solution state NMR is used to measure properties such as identification, quantitative composition, tacticity, monomer sequencing and defects (end groups, branching etc.) Much of the work reported arises from "in-house" examples, however, literature data are included for completeness where appropriate. "Tricks of the trade" for obtaining good quality NMR spectra of synthetic polymers are included where necessary.

1 Introduction

Nuclear Magnetic Resonance (NMR) Spectroscopy is well recognised as the primary technique for the determination of molecular structure of synthetic polymers. In this article the author aims to illustrate how solution-state NMR is used for molecular structural determinations on a wide range of commercially available polymers.

Bovey and co-workers in the 1960s pioneered the use of NMR for the study of polymer structure. Subsequent progress in magnet design, more sophisticated electronics on spectrometers and advanced computers has enabled NMR spectroscopists to delve deeper into the structure of polymers. Today all major polymer manufacturers have support research teams and generally these laboratories will have a heavy financial commitment in NMR spectroscopy.

This article will illustrate how solution-state NMR is used in an industrial laboratory with applications from the synthetic polymer area. Much of the work reported will be from the author's own studies, but in addition will include literature data where appropriate. In particular, literature references will be given on papers which provide a greater in-depth analysis than is possible in the space available in such an article.

Solution-state NMR data will be used to illustrate how NMR can be used to answer the questions frequently asked by polymer chemists:

(i) What is this polymer?
(ii) What is its composition?
(iii) How is it put together?
(iv) What are the irregularities, e.g. branches and end groups.

The form of the article is that each section will consider the application of NMR techniques to one class of polymer, e.g. polyolefins, condensation polyesters, etc. and all aspects of question i) to iv) will be addressed where necessary.

For NMR studies of polymer chemistry the NMR of ^{1}H and ^{13}C nuclei are the most valuable; other nuclei which are used less frequently are ^{19}F, ^{29}Si and ^{31}P. Clearly for solution-state NMR studies the polymers need to be dissolved and comprehensive consideration is given in the Experimental section to selection of suitable solvents, preferably deuterated, and preferred instrument conditions. Polymer solutions frequently require elevated temperatures to prevent precipitation or phase separation and most modern NMR spectrometers can examine solution up to 150 °C.

In our industrial polymer research laboratory we use NMR at all stages in the development of new or existing polymers. For instance, in the early stages of the development of a new polymer NMR would be used to check the structure and purity of monomers. Polymers produced in the initial polymerisations may well be low molecular weight and NMR would determine composition, end group types and structure of the polymer. As the programmes progressed with increases in molecular weight of the product and changes in molecular structure produced by process modifications these would be monitored by NMR. Often at this stage patent cover is sought and NMR data are frequently used to specify the molecular structure of the products. When polymer production progresses from Laboratory, Semi-Tech and Pilot Plant to Main Plant the use of NMR tends to diminish, although we find we have a steady demand for NMR work from polymer chemists who are effecting modifications to 'well-established' polymers.

In the vast majority of companies the high cost of modern high-resolution NMR spectrometers precludes their wide-scale use on production plants. However, low resolution, relatively low-cost ($\leqq$£20 K) NMR instruments are available, which find applications which involve the quantitative measurement of the mobile phase in a polymer. Applications include the determination of the concentrations of plasticiser in polyvinyl chloride, rubber in polymer blends, and 'atactic' polymer in polypropylene. Suitable instruments for this type of analysis are: Bruker's "Minispec" and Oxford Instrument's "Analysers". These instruments are robust, simple to use, reliable and can be modified for in-line monitoring.

Infrared spectroscopic test methods are widely used in industry for quality control testing of polymers. This technique offers advantages over high-resolution NMR including:

(i) Cost of instrumentation
(ii) Ease of operation
(iii) Easily achievable low measurement error
(iv) Short time for analysis

In our laboratory we frequently determine calibration standard polymers by ^{1}H- or ^{13}C-NMR which are then used to establish an infrared spectroscopic method. Intensities of bands in the infrared spectra of polymers are frequently dependent upon the monomer sequencing in the polymer. This in the majority of cases rules out the use of homopolymer blends to establish calibration samples for a copolymer method. NMR calibrated infrared test methods developed over the years in our laboratory include: ethylene in propylene/ethylene copolymers, block and random; vinyl acetate in ethylene/vinyl acetate, and vinyl chloride/vinyl acetate copolymers. This approach is frequently used but not reported widely in the literature. However, Fisch and Dannenberg [1] used ^{13}C-NMR on a series of propylene/butene copolymers to establish calibration standards for a hot-cell infrared method. Randall [2] reported a similar approach for propylene/ethylene random copolymers.

One aim of this article is to stress the applications of NMR to synthetic polymers, so intentionally details of the theoretical aspects will be kept to a minimum. The reader is referred to the following references for further reading: general solution-state NMR theory [3–5], books on NMR of polymers (6–7), and other review articles on polymers [8–9].

Reference spectral collections on NMR spectra of polymers are few in number and for ^{1}H-NMR the author recommends the Sadtler collection [10]. Although dated in the sense of being low field (60 MHz) spectra, the very wide range of spectra shown and the excellent text make this book very useful in an industrial polymer lab. The other collection recommended is by Pham [11–13]. References [6 and 10] both give details of the nomenclature used to describe homopolymers and copolymers.

In the space available in this article it is neither possible to review the work on all polymers nor give an in-depth report of the polymers considered. The work described will be biased towards the polymers with which the author is familiar but in general this range of polymers covers the commercially important polymers.

Polymer blends have over the last few years taken on an important role in filling gaps in product ranges. A section is given over to these products and will consider the use of NMR to the determination of chemical changes on blending, e.g. transesterification and amide/ester exchange. The determination of composition of blends by NMR is not considered in this article.

2 Obtaining an NMR Spectrum of a Polymer

To obtain spectra of synthetic polymers which are suitable for qualitative or quantitative analysis spectrometer frequencies of $\geqq 200$ MHz for ^{1}H-NMR are recommended. The early work in the 1960s used 60 MHz, 100 MHz or 220 MHz continuous wave spectrometers. In today's laboratories spectrometers of frequencies between 200 and 500 MHz are common and all of these operate in the pulsed, or Fourier transform, mode. In sensitivity, dispersive, ease of operation and data manipulation these spectrometers have considerably improved over spectrometers of say 15 years ago.

In this section we will consider the practical requirements needed to take suitable NMR spectra of polymers. Four general headings will be considered:

(i) Solvents
(ii) Instrument parameters
(iii) Data manipulation and computing
(iv) More advanced pulse experiments

2.1 *Solvents and Techniques for Dissolving Polymers*

Suitable solvents for solution-state NMR studies on synthetic polymers range from chloroform to strong acids such as conc. sulphuric and triflic acid. Where polymers dissolve in solvents which are available in the deuterated form, the use of deuterated solvent has two advantages:

(i) the field/frequency lock on most spectrometers is a deuterium NMR signal. Thus the solvent provides for this, and in addition the deuterium resonance is used for magnet field homogeneity adjustment either manually or on the most modern spectrometers computer controlled.

(ii) the spin–lattice relaxation times (T_1) of the nuclei in deuterated solvents are much longer than comparable protons or carbons in polymers. Thus for pulse sequences suitable for polymers this results in the resonances from the solvent being reduced in intensity and hence reduces interference in the spectrum. The choice of suitable solvents and typical temperatures for NMR examinations for the more common polymers are given in Table 1.

When a polymer does not dissolve in a solvent available in the deuterated form, protic solvents would be used. In this situation the solvent is chosen so that its resonances have minimum interference with those of the polymer. For instance, in this way olefinic polymers can be examined in aromatic solvents such as 1,2-dichlorobenzene and 1,2,4-trichlorobenzene. When these solvents are used the deuterium necessary for field/frequency locking and magnet homogeneity

Table 1. Selected NMR solvents for representative polymers

Polymer	Solvent	Temperature (°C)
Polyolefins	*ortho*-dichlorobenzene (*o*-$C_6H_4Cl_2$)	100–120
	D_2-1,1,2,2-tetrachloroethane (TCE)	110
Condensation polyesters	(i) copolymers chloroform-D_1 ($CDCl_3$)	ambient – 50
	TCE	110
	(ii) homopolymers TCE	110
Polyethers	(i) oxymethylenes TCE	110
	o-$C_6H_4Cl_2$	100–120
	(ii) other polyethers $CDCl_3$	ambient – 50
	D_6-benzene (C_6D_6)	ambient
Vinyl chloride	*o*-$C_6H_4Cl_2$	100–120
polymers & vinyl polymers	TCE	110
Nylons	methanesulphonic acid	ambient
	conc. sulphuric acid	ambient
	fluorosulphonic acid	ambient
	formic acid	ambient
Polyurethanes	deuterated dimethyl sulphoxide D_6-DMSO	ambient – 100
Polysulphones	D_6-DMSO	80
	TCE	100
Polyketones	conc. acids e.g. sulphuric or trifluoromethane sulphonic	ambient
Ethylene copolymers (not olefins)	*o*-$C_6H_4Cl_2$	100–120
	TCE	100
	C_6D_6	ambient
Polyacrylates	$CDCl_3$	ambient – 50
	TCE	110
Condensation formaldehyde polymers	D_6-DMSO	ambient – 80

adjustment must be provided separately. The two common methods employed are:

(i) addition of deuterated solvent to the polymer solution.
(ii) placing a deuterated solvent in a concentric tube.

Deuterobenzene has been favoured in the former method and although well above its boiling point has been used at 100°–120 °C in *ortho*-dichlorobenzene for polyolefins to provide a lock signal for overnight accumulating. However, as benzene is a carcinogen we have severely restricted its use in our laboratory. Deuterated dimethyl sulphoxide has found favour when employed in method (ii) for carbon-13 studies. This is because d_6-DMSO has a strong deuterium resonance, has a high boiling point and is fairly stable for high temperature operation. For ^{13}C-NMR studies the polymer solution is contained in a 10-mm or 12-mm OD NMR tube and the deuterated solvent placed in a 5-mm OD NMR tube. When the 5-mm and 10-mm tubes are the same length and plastic caps are in place the 5-mm tube sits concentrically inside the wide tube.

^{1}H-NMR of polymer in protic solvents presents two difficulties:

(i) dynamic range of resonances from solvent and polymer
(ii) lack of deuterium resonance for field/frequency lock up and magnet field homogeneity shimming

To overcome the lack of deuterium resonance two-methods are employed. Firstly, we reverse the capillary method used for ^{13}C-NMR, that is the polymer solution is contained in the smaller tube and the deuterated solvent in the larger tube. This method is preferred for samples examined between ambient and 80 °C. Secondly, a deuterated solvent may be added to the polymer solution. This method is not used very often because of the possibility of precipitation of the polymer solution, incompatibility of solvents and the need for a fairly high concentration of deuterated solvent needed for sufficient deuterium signal.

In general, ^{13}C-NMR spectra can be taken on more viscous solution than ^{1}H-NMR. Where high signal-to-noise is required for ^{13}C-NMR spectra the concentration of polymer in solution should be maximised. However, if high-resolution spectra are needed for such as monomer sequencing determinations then more dilute solutions are used, typically ca. 10%.

In ^{1}H-NMR studies of synthetic polymers the same principles apply as for ^{13}C-NMR except for ^{1}H-NMR the viscosity of the solutions in all cases needs to be much lower. A simple test we use frequently in our laboratory is to heat the polymer solution in an NMR tube or test tube to the intended temperature for the NMR experiment, and the tube then is tilted. If the polymer solution does not flow freely along the tube the solution is too viscous and need to be diluted.

Semi-crystalline polymers do not dissolve readily and to facilitate dissolution we prepare amorphous samples. This we achieved by hot pressing the polymer to just above its melting temperature between polytetrafluoroethylene (PTFE)-coated stainless-steel plates and then rapid cooling into a bucket containing iced water. This approach we have used very successfully for some polyesters, e.g.

polyethyleneterephthalate, polyethylene naphthoate and some liquid crystal polyesters.

Many polymers are either only soluble at elevated temperatures or require heating to dissolve and for this purpose we have thermostatically controlled heating blocks. In practice we have two blocks, with temperature ranges ambient to 80 °C and 80 °C to 150 °C. Polymers are dissolved in test tubes for ^{1}H-NMR examination and transferred by glass pasteur pipettes to the NMR tube. For ^{13}C-NMR studies we use 10-mm tubes for the majority of our work and where either temperatures above 80 °C are required to dissolve the polymer or concentrated acids are needed as solvent, the polymers are dissolved in the NMR tube. Where solutions are prepared in test tubes or 10/12-mm NMR tubes we find a length of 3/4-mm stainless steel rod operates as a very useful tool for stirring solutions.

2.2 *Instrument Parameters*

The length of this article does not allow the author to consider all the parameters which can be changed on a modern NMR instrument. However, a few parameters will be considered which affect either the quality of spectra or the ability to achieve quantitative data. In the following discussion the instruments are assumed to be in pulsed mode operation. All modern spectrometers with proton frequencies $\geqq$ 200 MHz operate in this mode.

The role of the computer in spectrometer control has increased significantly in the last few years. The early pulsed spectrometers had a small computer for spectrometer control and data storage but many of the parameters had to be set normally. Shimming of the magnet homogeneity on those spectrometers was considered by many operators to be an art rather than a science.

On currently available pulsed spectrometers the computers are considerably more powerful and the vast majority of the instrument parameters are computer controlled. Operation of spectrometers under computer control has led to the use of automatic sample changers with the result that expensive spectrometers can be used 24 hrs per day, 7 days per week. In the author's laboratory we estimate that use of a sample changer at least doubles and nearly triples the output of a spectrometer. This is based on a work load requiring both proton and carbon-13 spectra.

Operation and use of pulsed NMR instruments is considered in great detail elsewhere and the author refers the reader to three books. A recent book edited by Field and Sternhall [14] considers numerous examples of the use of NMR in analysis; a practical approach in the use of Fourier transform NMR spectrometers is considered by Müllén and Pregosin [15] and Shaw's book on Fourier transform NMR has recently been republished [16].

Pulsed NMR spectrometers employ a single high-power, short-duration pulse of r.f. radiation to excite simultaneously all nuclei close to the frequency of the pulse. The signal from the excited nuclei is monitored with time immediately following the pulse and this response vs time is collected by the computer is known as the free induction decay or FID. The time for which the FID is sampled is known as the acquisition time (AT) and the interval between pulses is determined by the sum of the AT and a variable delay known as the pulse delay (PD). The time the high-power r.f. pulse is applied is defined as either the pulse width (PW) or pulse angle. The 90° pulse is the power required to obtain a maximum response from the sample. The pulse width required to achieve a 90° pulse is dependant upon the power of the r.f. pulse.

The rate at which the FID is sampled is determined by the speed of the analogue-to-digital converter (ADC). The interval (Δt) between points along the FID defines the detected spectral width (SW) of the FT NMR experiment.

$$SW = \frac{1}{2\Delta t}$$

The acquisition time used in an experiment may limit the resolution achieved in the spectrum. This is because it can be shown theoretically that to resolve two resonances separated by $\Delta \nu$ Hz the FID must be acquired for at least $1/\Delta \nu$ seconds. Thus, if the AT is 0.5 s the resolution achievable will be 2 Hz.

Mathematically: $AT = N/(2 \times SW)$

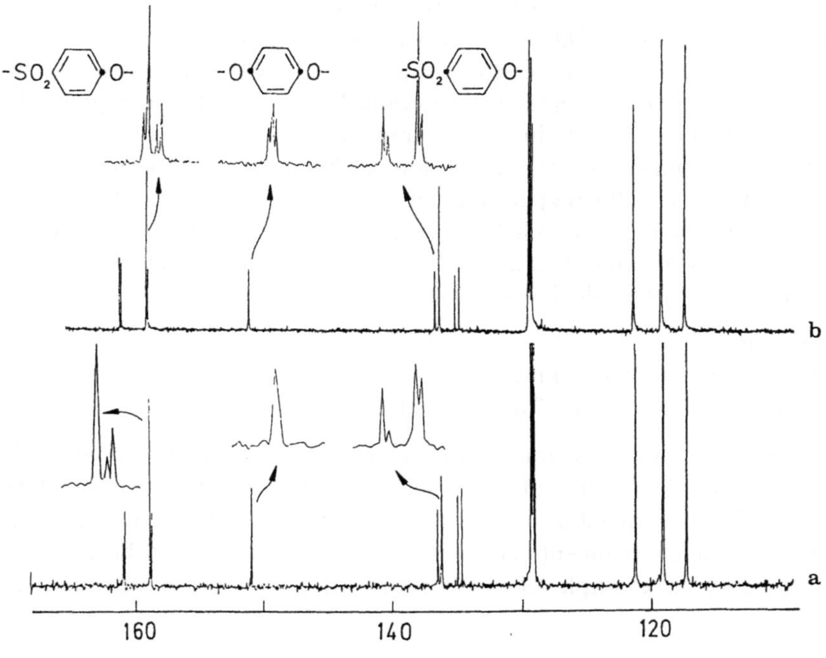

Fig. 1 a, b. Partial ^{13}C-NMR spectrum of polyarylene ether sulphone/ether ether sulphone copolymer. (**a**) 236-ppm spectrum acquired AT = 0.51 s. (**b**) 60-ppm spectrum acquired AT = 2.0 s

where N is the number of data points in the FID. Thus, to increase AT it is usual to reduce SW. The improvement in resolution which can be achieved is illustrated in Fig. 1. The spectrum is a part spectrum of a polyarylene ether sulphone copolymer and resonances from the two types of quaternary carbons are illustrated. The top spectrum was taken with a 236-ppm spectral width and an AT of 0.51 s and the lower was 60-ppm SW and an AT 2.05 s. The improvement in resolution between the two spectra can clearly be seen.

For quantitative experiments the magnetisation after application of the pulse must return to equilibrium and for this a pulse delay of $5 \times T_1$ (T_1 is the spin-lattice relaxation time.) must be allowed between pulses. Fortunately in ^{13}C-NMR, T_1 values for chain methine and methylene carbons in polymers are short ($\leqq 250$ ms) [17] and relatively short PD values are required. However, for carbons in branches [18] or quartenary carbons [17] T_1 values of up to a few seconds may be encountered.

2.3 *Data Manipulation and Computing*

Manipulation of the data in this context refers to the massaging of the FID prior to Fourier transforming to obtain the normal spectrum and methods for integration of resonances for quantitative measurements. The most useful and widely used message of FID either improves signal-to-noise (S/N) or improves resolution. Enhancement of S/N is achieved by weighting the FID in favour of the data at the beginning. This is typically achieved by multiplying the data of the FID by a function which decreases exponentially with time e.g. e^{-at}. This process improves S/N but degrades resolution. Multiplying by an exponential function which increases with time e.g. e^{at} improves resolution but loses S/N.

Many massage functions are available for weighting FIDs and several of these methods are found in software packages on modern spectrometers. The effect of massaging of a polymer ^{13}C-NMR spectrum with similar functions to those described above is shown in Fig. 2. The 67.8 MHz ^{13}C-NMR spectra shown are of an propylene/ethylene block copolymer:

(a) no massaging
(b) line broadening with e^{-at} function
(c) resolution enhanced with e^{at} and linear function.

Integration of resonances for quantitative analysis is in most spectra perfectly adequately achieved with the software packages available on the spectrometers. When severely overlapped resonances require integration, curve fitting is preferable and some instrument manufacturers supply software for this analysis. Figure 3 shows an example of curve fitting of a complex spectrum with

Fig. 3 a–d. Carbonyl resonances of a copolymer of methyl methacrylate and methacrylic acid. (a) original spectrum, (b) curve fitted spectrum (c) individual components of curve fitting, (d) residue between a & b. Spectrum taken in d_6-DMSO at 80 °C at carbon-13 operating frequency of 100 MHz

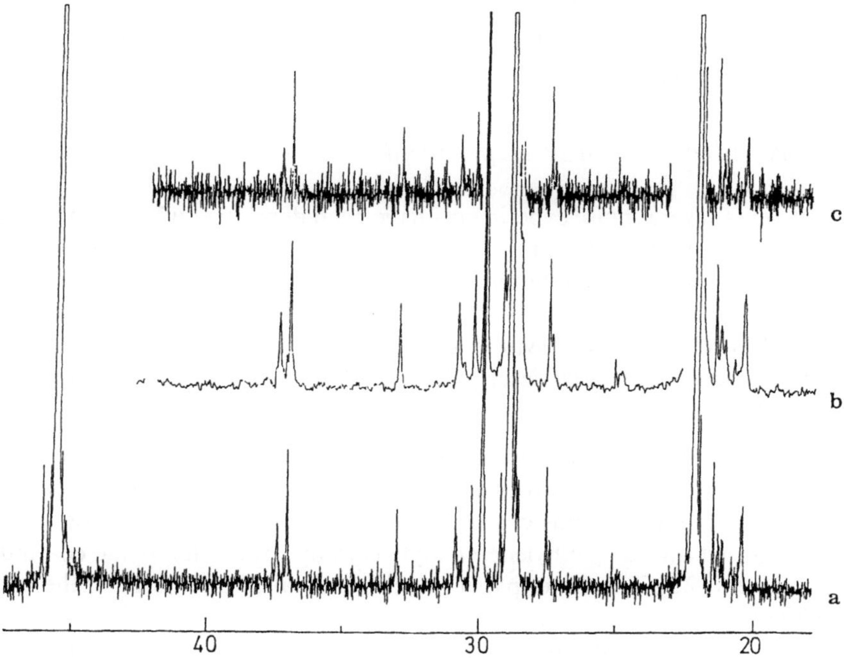

Fig. 2 a–c. ^{13}C-NMR spectrum of a propylene ethylene block copolymer. (**a**) no massaging of data (**b**) line broadened by e^{-at} function. (**c**) resolution enhanced with combined e^{at} and linear function

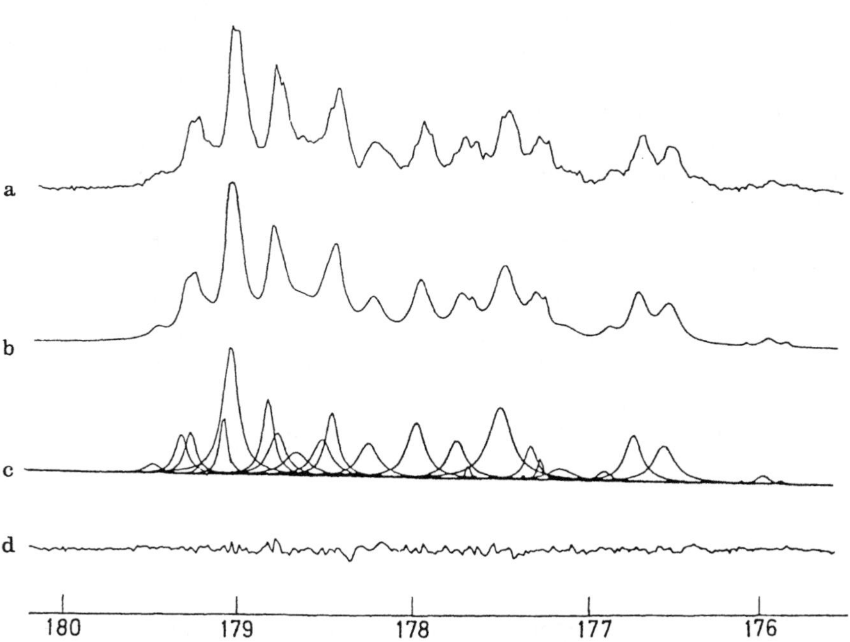

commercial software [19]. The spectrum is the carbonyl region of a ^{13}C-NMR spectrum of a methyl methacrylate/methacrylic acid copolymer and the levels of residuals indicate a good fit.

2.4 More Advanced Pulse Experiments

In recent years several pulse experiments have been developed which will either simplify spectra by spectral editing or lead to an increase in information relating one resonance to another. The latter experiments are known as two-dimensional (2-D) NMR.

In the 2-D NMR study of synthetic polymers the two commonly used 2-D NMR experiments, with brief descriptions and literature references for further reading, are:

(i) ^{1}H–^{1}H chemical shift correlations, J-correlated spectroscopy (COSY). This experiment provides a map of all homonuclear couplings, and thus the connected coupled protons can be deduced, as reported for cellulose [20], polyvinyl butyral [21], ethylene/methyl acrylate/carbon monoxide co- and terpolymers [22]. This latter paper presents a detailed discussion of the use of COSY.

(ii) ^{1}H–^{13}C chemical shift correlation. In most applications the ^{13}C–^{1}H correlation maps the chemical shifts of protons to the carbon-13 signals of the corresponding directly attached carbon. This is particularly useful when either ^{1}H- or ^{13}C-NMR spectra are well assigned. In addition, possible configurational sequence assignments can be deduced from the correlation of carbons and protons resonance of methylene groups. This arises because when a single carbon methylene shift correlates with a single proton shift, then the protons in the methylene group are equivalent and this indicates a racemic unit. Whereas a simple carbon shift correlates to two proton chemical shifts, this indicates non-equivalent protons in the methylene group, i.e. a meso (m) unit. Literature references for the use of this technique include polyvinyl chloride [23] and polymethyl methacrylate [24].

In ^{13}C-NMR spectra of polymers it is often very useful to know whether a resonance arises from a methyl, methylene, methine, or quaternary carbon. Pulse sequences determine multiplicities and spectra editing are APT (Attached Proton Test) [25] and (Distortionless Enhancement by Polarisation Transfer) DEPT [26]. Heatley in his report of ^{13}C-NMR spectra of ethylene oxide/propylene oxide [27] illustrates the use of these techniques and shows examples whereby triad sequences can be determined.

3 Polyolefins

Olefin homopolymers and copolymers are commercially very important and the polymers based on ethylene and propylene provide the largest capacity. Polymers in this category are:

(i) polyethylene
(ii) polypropylene
(iii) ethylene/propylene (E/P) copolymers including:
  - P/E random and block copolymers,
  - P/E rubbers.

Other olefin copolymers considered in this section are ethylene/α-olefin copolymers, propylene/butene copolymers, and propylene/butene/ethylene terpolymers.

In addition, the NMR spectra of the commonly encountered ethylene, non-olefine copolymers are discussed e.g. ethylene copolymers with vinyl acetate, ethyl acrylate, etc.

The ^{1}H-NMR spectra of polyolefins even at high frequencies are complex and the preferred NMR examination is ^{13}C-NMR. Suitable solvents systems for these polymers are: deuterated 1,1,2,2-tetrachloroethane, 1,2-dichlorobenzene/deuterobenzene (C_6D_6), and 1,2,4-trichlorobenzene/C_6D_6. Spectra are taken with the solution maintained at a controlled temperature between 100 °C and 130 °C, usually 100 °C.

3.1 Homopolymers

3.1.1 Polyethylene

Solution-state carbon-13-NMR has been used extensively to study the molecular structure of low-density polyethylene (LDPE). This polymer is manufactured at high pressures and temperatures and as early as 1940 Fox and Martin [28], in an infrared spectroscopic study, showed that the concentration of methyl groups was higher than could be accounted for by end groups. Thus they concluded that methyl-containing branches were present in LDPE. In the 1950s further infrared spectroscopic studies [29, 30] dealt with the nature and concentration of these branches. However, a considerable amount of information on branch types and concentrations has been gained by carbon-13-NMR spectroscopic studies of LDPE.

The branches in LDPE are divided into two categories, short-chain branching structures formed by intramolecular hydrogen transfer reactions and long-chain branching formed by intermolecular hydrogen transfer reactions. The nature and concentration of short-chain branching has received considerable attention by several groups of workers [31–37]. Figure 4, show a 67.8 MHz ^{13}C-NMR spectrum of a typical LDPE and full analysis of this spectrum shows the presence of the branching types shown below. To assign resonances in spectra of this type, empirical methods of calculating chemical shift values for hydrocarbons [38–39] and the use of spectra of ethylene/olefin copolymers [31, 40] with known branch types have been of considerable value.

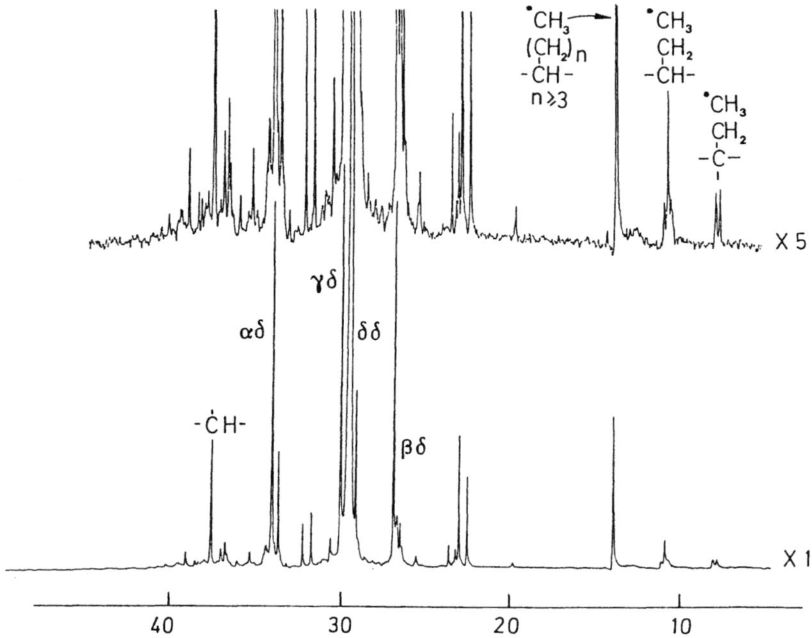

Fig. 4. 67.8 MHz ^{13}C-NMR spectrum of low-density polythene, solvent deuterated 1,1,2,2-tetrachloroethane and spectrum taken at 110 °C. Pulse delay 5 s

Branches found in LDPE:

$$CH_3(CH_2)_n\overset{|}{\underset{|}{C}}H \qquad n = 0, 1, 2, 3, 4, 6+$$

$n = 0$ if propylene modified; $n = 2$ rarely present.

$$\text{—}\underset{\displaystyle CH_2\text{—}CH_3}{\underset{|}{CH}}\text{—}CH_2\text{—}\underset{\displaystyle CH_2\text{—}CH_3}{\underset{|}{CH}}\text{—}CH_2\text{—}, \quad \text{—}CH_2CH_2\underset{\displaystyle CH_2\text{—}CH_3}{\underset{|}{CH}}CH_2CH_2\text{—}$$

and

$$\text{—}\overset{|}{\underset{\displaystyle \overset{\displaystyle CH_2}{CH_3}}{\underset{|}{C}}}\text{—}$$

The identification of the types of branches in LDPEs has proved to be less controversial than attempts at quantitative measurements of the concentration of these branches. The need for long delays between pulses (at least 30 s) before quantitative data were achieved was demonstrated [18] as was the fortunate fact that the maximum nuclear Overhauser enhancement (NOE) factor of 2 was

Table 2. Branch concentration of low density polyethylenes prepared with a range of polymerisation conditions

Polymerisation pressure kg/cm^2	Polymerisation temperature °C	CH_3/1000C**	Et/1000C	Bu/1000C	$\geqq C_3$/1000C
2500	320	34	ND*	16	16
	240	16	ND*	8	8
	140	8	ND*	4	4
2000	320	40	9	17	14
	240	20	3	10	7
	140	9	ND*	5	4
1500	300	44	9	20	15
	140	10	ND*	5	5
1400	320	45	14	17	14
	300	39	11	15	13
	180	17	ND*	10	7
1300	320	61	20	20	21
	220	24	5	10	9
	180	19	~1	9	9
1200	320	60	21	22	17
	300	52	18	21	13
	220	24	4	11	9
	140	10	ND*	5	5

* none detected, that is <1
** measured by IR test method (Willbourn AH, Ref. [30])

experienced for all carbon resonances in the spectra of LDPEs. Thus, for these systems, it would seem that errors in intensity measurements caused by differential NOE factors can be neglected.

The effect of temperature and pressure on the concentration and type of branching in LDPE has been studied in our own laboratory. Table 2 shows some typical data, and the procedure for the determination of ethyl, butyl and $>$ butyl branches was as described in Ref. [31].

These data show some interesting trends and we concluded that:

(i) the activation energy for formation of butyl and $>C_4$ is about the same at all pressures, and appreciably less than that for ethyl groups;
(ii) at a constant temperature of 320 °C the formation of ethyl groups increases rapidly with decreasing pressure, whereas for butyl and $>C_4$ there is very little dependence on reactor pressure;
(iii) for most of the polymers the ratio of butyl to $>$butyl is roughly equal to unity.

3.1.2 Polypropylene

Polypropylene is manufactured by the polymerisation of high-purity propylene with a Ziegler–Natta catalyst. These heterogeneous catalysts are designed to produce highly isotactic semi-crystalline polymer, reproducibly and with high yields. As a by-product a low percentage of non-stereo regular ('atactic') polymer is also produced and in some polymerisation vessels this is removed into the hydrocarbon used for the polymerisation.

Polypropylene is used in a wide range of applications as a moulding polymer and an important area commercially is in its use as an oriented film. The stiffness of polypropylene is improved by fillers-talc and calcium carbonate are commonly used at high levels (ca. 40%).

In addition to isotactic polypropylene, the syndiotactic form is semi-crystalline and early ^{1}H-NMR studies by Bovey and co-workers paved the way for the use of this technique in determination of tacticity. In the ^{1}H-NMR spectrum of syndiotactic polypropylene the protons of the methylene group are equivalent, whereas there is a non-equivalence in the isotactic polypropylene case. This information has been used and developed to assign ^{1}H-NMR spectra of homopolymers and recently in the 2-D-NMR study of PVC by Levy and co-workers [23] it was used as a basis for the ^{13}C-NMR assignments.

Early NMR studies used ^{1}H-NMR and several reports of tacticity measurements by ^{1}H-NMR appeared [41–44]. In our laboratory, for several years we used the method of Heatley and co-workers [41]. In this method the methylene resonances of the racemic dyads were compared with a ^{13}C satellite and thus quantified. Although very useful at the time, subsequent reports showed that the resonance being measured was characteristic of only -rrr- and -mrm- sequences and mrr sequences were not determined. As it is now generally agreed that the major defect structure in isotactic polypropylene is the -mmrrmm- sequence [45], Heatley's method underestimated the total racemic dyad concentration and in reality the measurement was biased in favour of the long *r* sequences in the 'atactic'.

The ^{13}C-NMR spectra of polypropylenes are more informative than the corresponding ^{1}H-NMR spectra. All the resonances in the ^{13}C-NMR spectra of polypropylene show a structure which indicates tacticity information. The detailed discussion of the assignments of these resonances will not be discussed here but the reader is referred to literature papers [45–47]. The resonances from the α-methyl groups can readily be resolved into pentad sequences [46]. Figure 5 shows typical spectra of a highly isotactic and a polymer which contains a relatively high level of 'atactic', and the assignments are shown in this diagram. Some of the resonances show further splitting and the detailed assignments of these are discussed further [46, 47]. The relative proportions of pentad sequences can be determined from these α-methyl resonances and these data are used industrially to monitor effects of polymerisation and catalyst changes. ^{13}C-NMR spectroscopic data have confirmed a two-site model for polypropylene polymerisation [48]. In simple terms this means that different catalyst sites produce either high isotactic polymer or stereo-irregular polymer (referred to as 'atactic' although in reality it is syndiotactic-rich). The latter is soluble in hydrocarbon solvents and may be extracted.

Polypropylene prepared with vanadium-based catalysts contain propylene unit inversions and the concentration of such units can be determined by ^{13}C-

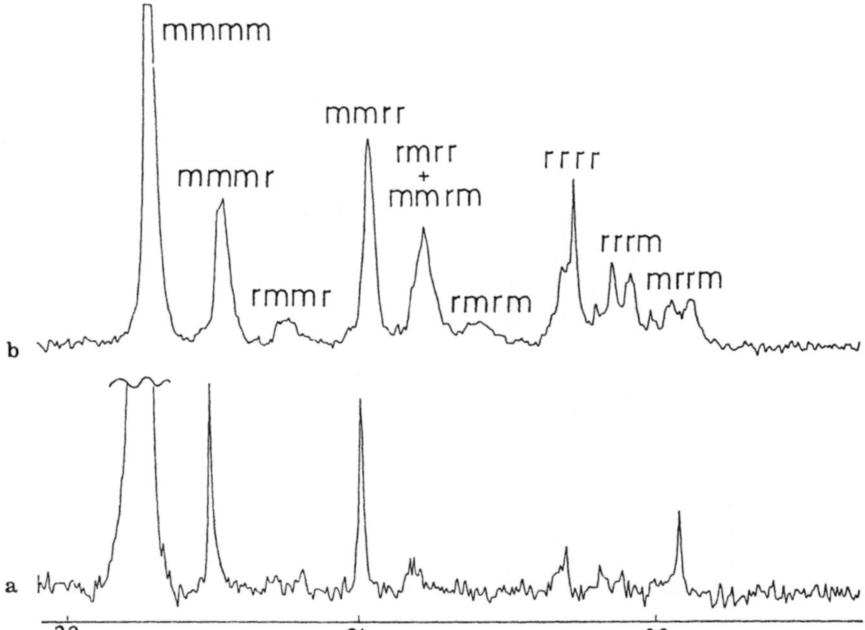

Fig. 5a, b. Resonances from the α-methyl carbons in ^{13}C-NMR spectra of: (**a**) highly isotactic polypropylene, (**b**) low stereoregular polypropylene

NMR [49, 50]. Most use is made of vanadium catalysis in the preparation of copolymers and terpolymers of propylene as thermoplastic rubbers (see Sect. 3.2).

3.2 Olefin Copolymers

Ethylene/α-olefin copolymers, known as linear low-density polyethylene (LLDPEs) has become important commercially over the last few years and Fig. 6 shows ^{13}C-NMR spectra of typical polymers:

(a) ethylene/butene-1 and
(b) ethylene/octene-1.

^{13}C-NMR spectroscopy is an ideal method for the identification of LLDPEs and other reference spectra for other ethylene α-olefin copolymers are to be found in the literature [40, 51].

Propylene copolymers and terpolymers with ethylene and butene are available commercially in various types. Figure 7 shows ^{13}C-NMR spectra of two types:

(a) propylene/butene copolymer (~20% wt butene) and
(b) a propylene/ethylene/butene terpolymer.

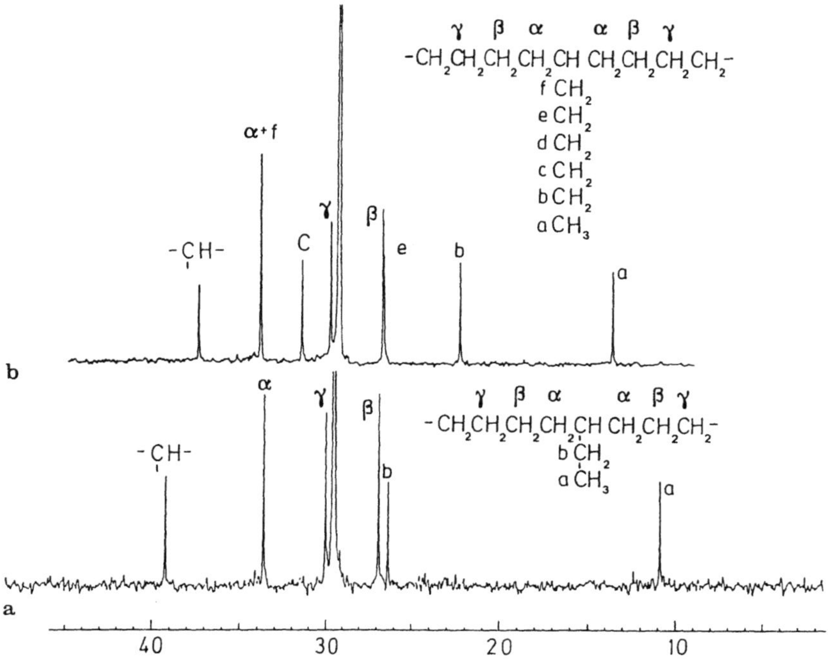

Fig. 6 a, b. 67.8 MHz ^{13}C-NMR spectra of: (**a**) ethylene/butene-1 copolymer, (**b**) ethylene/octene-1 copolymer

The assignments of the resonances are given on the figure and for the terpolymer it can be demonstrated that each butene and each ethylene is flanked by two propylene units. Detailed discussion of these assignments are reported elsewhere [52–54], and for propylene/butene and for propylene/ethylene [55].

Propylene/ethylene copolymers are commercially very important and block copolymers with up to 30 wt% ethylene are used widely as thermoplastics. Similar physical properties to those for block copolymers can be achieved from a blend of polypropylene and an ethylene/propylene rubber. The latter often contain small quantities of copolymerised diene monomer and for this reason they are known as EPDMs (*e*thylene/*p*ropylene/*d*iene *m*onomer). Figure 8 shows a typical ^{13}C-NMR spectrum of a polypropylene/EPDM blend. There are similarities between this spectrum and the spectrum of propylene ethylene block copolymer in Fig. 2. However, resonances unique to the EPDM phase are highlighted with *'*s* and these arise from carbons in structures formed when propylene inversion occurs during polymerisation, so that the following sequences are formed:

$$-\underset{|}{CH}-CH_2-CH_2-\underset{|}{CH}-$$

and

$$-\underset{|}{CH}-CH_2-CH_2-CH_2-CH_2-\underset{|}{CH}-$$

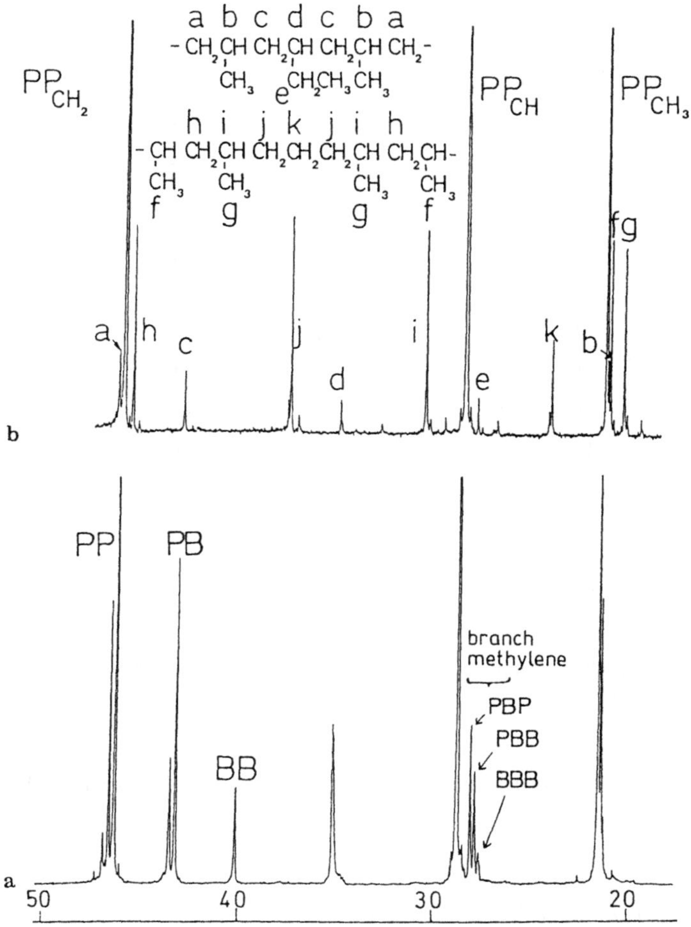

Fig. 7 a, b. 67.8 MHz ^{13}C-NMR spectra of (**a**) a propylene/butene copolymer (~ 80:20 w/w) (**b**) a propylene/butene/ethylene terpolymer with ca 2.5 wt% butene and 5.3 wt% ethylene

In Fig. 8 resonances from the EPDM are assigned using the nomenclature system proposed by Carmen and Wilkes [56]. In this, each carbon is designated two Greek letters depending upon the distance from a methine group. Thus we have:

$$-\underset{|}{CH}-\overset{\alpha\gamma}{CH_2}-\overset{\beta\beta}{CH_2}-\overset{\alpha\gamma}{CH_2}-\underset{|}{CH}-$$

The composition of unblended EPDM polymers can be determined from the ^{13}C-NMR data [55], and the author prefers to use the methylene resonances for this quantitative analysis. In block copolymers or polypropylene/EPDM blends the above analysis is more difficult because of severe overlap of the $\alpha\alpha$ carbon

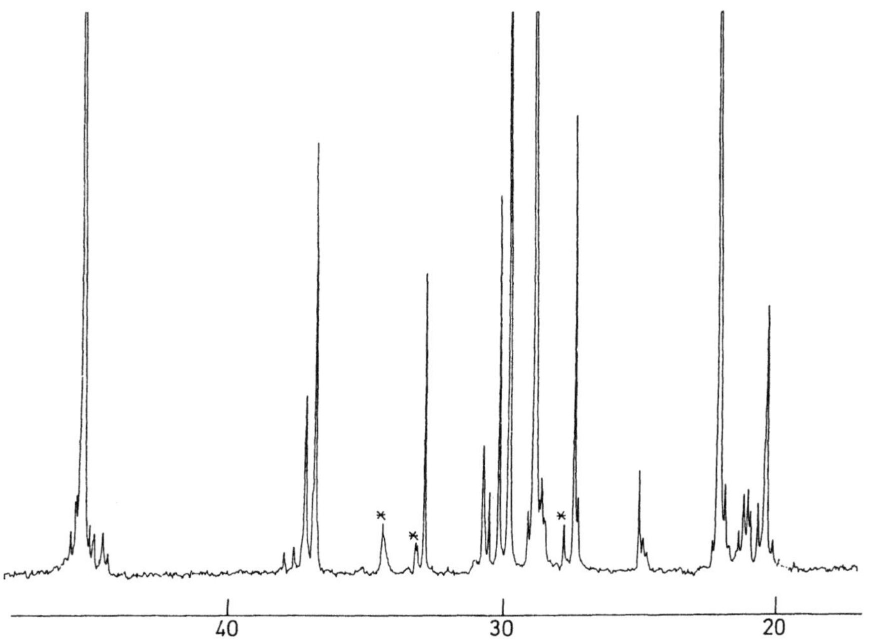

Fig. 8. 67.8 MHz ^{13}C-NMR spectrum of a polypropylene/EPDM blend. Peaks marked *arise from the presence of inverted propylene units

resonances of the copolymer by the corresponding resonances from the polypropylene.

3.3 Ethylene Copolymers (Non-Olefin)

Ethylene copolymers with vinyl acetate, methyl and ethyl acrylate find numerous applications as films, elastomers, and in moulding/extrusion areas. In routine identification for type and composition an ^{1}H-NMR spectrum taken in 1,1,2,2-perdeuterotetra-chloroethane suffices. The spectra (see Fig. 9) are dominated by the 1.25-ppm resonances. However, the protons adjacent to the ester groups are characteristic and these resonances may be integrated for quantitative analysis.

Commercially available ethylene/vinyl acetate (EVA) copolymers contain 5 to 50 wt% of VA. The position of the resonance from the VA methine is sensitive to environment, and triad sequences can be determined. This in itself is not particularly useful but it does allow polyethylene/EVA blends to be characterised for EVA type as the methine resonances "fingerprint" the composition of the EVA component.

To illustrate the methodology, consider the ethylene/vinyl acetate copolymer ^{1}H-NMR spectrum in Fig. 9a. The assignments are given in this figure and we denote the integrals of the acetate methyl group and 0.9-ppm to 1.75-ppm region

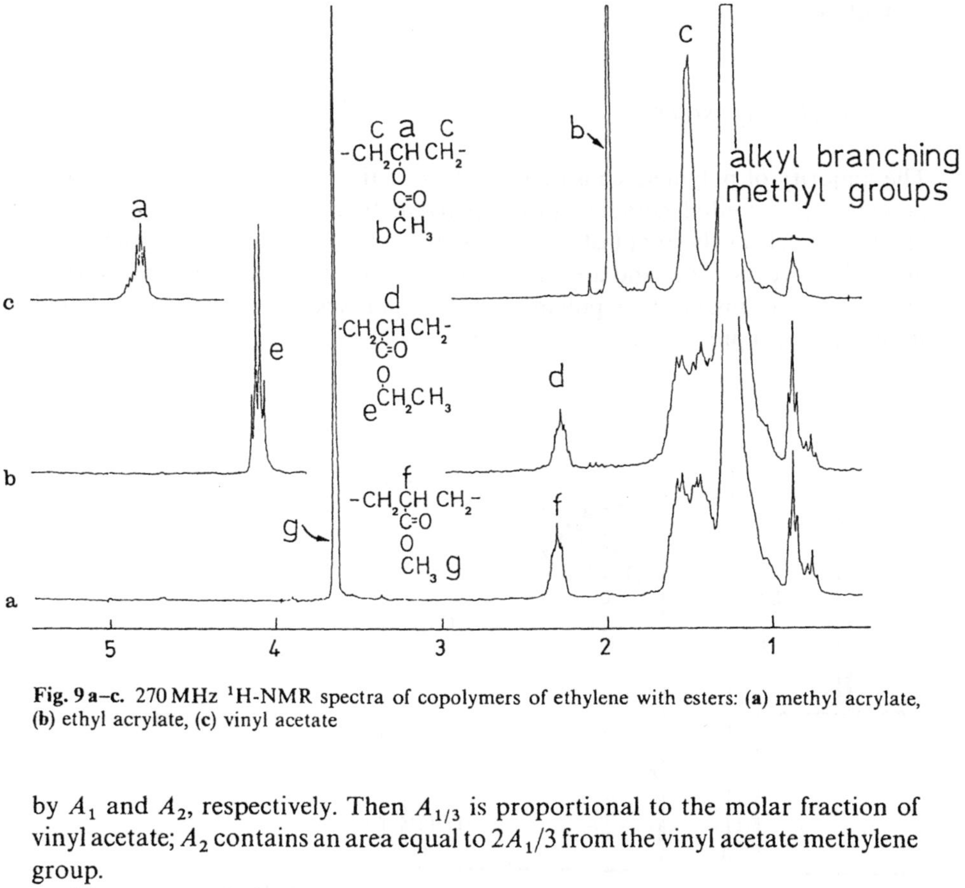

Fig. 9 a–c. 270 MHz ^{1}H-NMR spectra of copolymers of ethylene with esters: (**a**) methyl acrylate, (**b**) ethyl acrylate, (**c**) vinyl acetate

by A_1 and A_2, respectively. Then $A_{1/3}$ is proportional to the molar fraction of vinyl acetate; A_2 contains an area equal to $2A_1/3$ from the vinyl acetate methylene group.

Thus we conclude that:

$$\frac{\text{moles vinyl acetate}}{\text{moles ethylene}} = \frac{\left(\dfrac{A_1}{3}\right)}{A_2 - \dfrac{2}{3}A_1} \cdot 1/4 = x$$

$$\text{wt}\frac{\text{VA}}{\text{E}} = \frac{x \cdot 86}{28} = y$$

$$\text{wt\% VA in polymer}\,\frac{100y}{1+y}$$

^{13}C-NMR studies on ethylene/vinyl acetate have been found to be useful in MSD determination and determination of types and levels of alkyl branching. Work of this type has been reported [58]. Other literature references on copolymers not discussed in this article are: ethylene/vinyl alcohol [59, 60] and ethylene/vinyl chloride [61].

4 Polyesters

4.1 Vinyl Polyesters

The majority of polymers considered in this section are methacrylate/acrylate polymers. Polymethyl methacrylate or methyl methacrylate/acrylate copolymers are used widely as thermoplastics for mouldings. Copolymers with higher levels of acrylate are less rigid and are used as impact modifiers for polyvinyl chloride or as coating resins. Often in polymers of the latter type styrene and/or α-methyl styrene are also present as a comonomer.

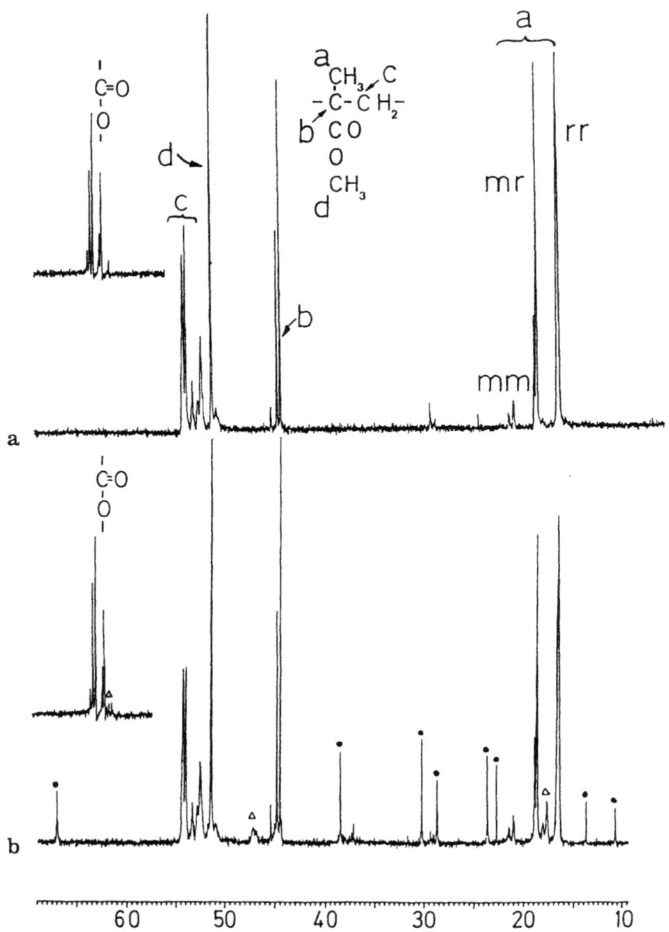

Fig. 10 a, b. 67.8 MHz ^{13}C-NMR spectra of: (**a**) poly methyl methacrylate (PMMA), (**b**) MMA/2-ethyl hexyl acrylate copolymer

The high methyl methacrylate polymers are readily soluble in deuterated forms of chloroform or 1,1,2,2-tetrachloroethane. Frequently encountered acrylate comonomers are methyl, ethyl, *n*-butyl, cyclohexyl, and 2-ethyl hexyl esters. ^{13}C-NMR spectroscopy provides the diagnostic evidence for acrylate type, however, in most cases ^{1}H-NMR spectroscopy is most suitable for quantitative determinations. Figure 10 shows ^{13}C-NMR spectra of:

(a) polymethyl methacrylate (PMMA) and
(b) MMA/2-ethyl hexyl acrylate.

The ^{13}C-NMR spectrum of polymethyl methacrylate (Fig. 10a) shows well-resolved resonances, and several literature reports have considered such spectra in detail [62–65]. The resonances from the α-methyl group carbons show triad sequencing information and proportions of triads for PMMA are usually close to the following:

isotatic triad mm	~6%
heterotactic triad mr	~39%
syndiotactic triad rr	~55%

The assignments of these and other resonances are shown in Fig. 10a. The many resonances from the carbonyl groups were most elegantly assigned to pentads by Moad and co-workers [24].

The vast majority of NMR studies on polyacrylates in our laboratory requires qualitative and quantitative analysis for comomer(s) and MSD determinations; tacticity determinations are not important because in general the tacticity does not change significantly from the values given above.

Comparison of the spectra of Fig. 10b and the PMMA spectrum of Fig. 10a show several difference and these can be classified into two types:

(i) resonances associated with the effect of copolymerising the 2-ethyl hexyl acrylate comonomer into a preliminary PMMA backbone. These resonances are highlighted with a Δ, and
(ii) resonances arising from alkyl ester carbons of the 2-ethyl hexyl acrylate comomer (marked with •'s).

In MMA/methyl acrylate copolymers the resonances from the effects of the acrylate portion will be seen but the carbon from the methyl ester is coincident with that of the PMMA. The most obvious resonance from copolymerised acrylate arises from the carbon adjacent to the ester group and characteristic chemical shifts for acrylates are:

methyl	55 ppm
ethyl	63 ppm
n-butyl	66 ppm
2-ethyl hexyl	67 ppm
cyclohexyl	74 ppm

Copolymers with higher levels of the acrylate component are often encountered and these cause evaluation difficulties in three ways:

(i) polymers are not particularly soluble and frequently form a gel with deuterochloroform; however, on gels reasonable ^{13}C-NMR spectra can be taken especially if the observation temperature is raised to 50 °C.
(ii) these polymers often include other copolymerised monomers such as styrene, α-methyl styrene, acrylic and methacrylic acid. When significant levels of copolymerised acids are present, d_6-dimethyl sulphoxide proves to be a more efficient solvent.
(iii) an additional complication in analysis arises because many of these types of polymers are multi-stage polymers. This is where the polymer is prepared in two or more distinct stages. To evaluate spectra of this type, a good set of reference spectra of known polymers is invaluable.

4.2 Condensation Polyesters

Condensation polyesters are prepared by polymerisation of glycol(s) and acid(s) and may be encountered as homopolymers, e.g. polyethylene terephthalate (PET), or copolyesters where the number of glycols and acids may be many. Fortunately for the majority of polyesters, ^{1}H-NMR spectroscopy provides sufficient information such that both qualitative and quantitative compositions

Table 3a. Glycols frequently encountered in polyesters

Name	Formula	Abbreviation
Ethylene glycol	$HOCH_2CH_2OH$	EG
Diethylene glycol	$HOCH_2CH_2OCH_2CH_2OH$	DEG
Triethylene glycol	$H(OCH_2CH_2)_3OH$	TEG
1,2-Propylene glycol	$HO{-}CH(CH_3)CH_2OH$	PG
1,3-Butanediol	$HO{-}CH_2CH_2CH(CH_3){-}OH$	1,3BD
1,4-Butanediol	$HO(CH_2)_4OH$	1,4BD
1,6-Hexanediol	$HO(CH_2)_6OH$	1,6HD
Neopentyl glycol	$HO{-}CH_2{-}C(CH_3)_2{-}CH_2{-}OH$	NGP
Cyclohexane dimethylol	$HO{-}CH_2{-}C_6H_{10}{-}CH_2{-}OH$	CHDM
1,3-Propane diol	$HO{-}CH_2CH_2CH_2{-}OH$	
Hydroquinone	$HO{-}C_6H_4{-}OH$	HQ

Table 3b. Acids frequently encountered in polyesters

Name	Formula	Abbreviation
Terephthalic	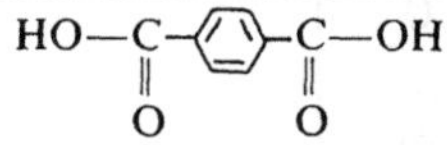	TA
Isophthalic	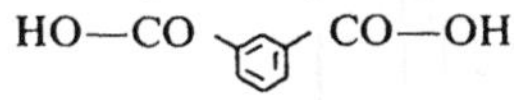	IPA
Orthophthalic	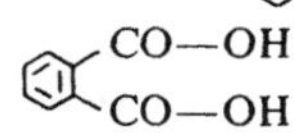	
Naphthoic	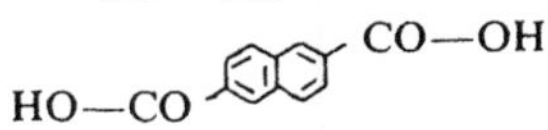	N
Aliphatic dicarboxylic acids	HO—CO$(CH_2)_n$CO—OH $n = 2$ succinic $n = 4$ adipic $n = 6$ suberic $n = 7$ azelaic $n = 8$ sebacic	
Fumaric/maleic	HO—COCH=CH—CO—OH	

can be determined, end groups concentrations measured and, in some copolyesters monomer sequencing determinations made. However, in general, ^{13}C-NMR spectroscopy is preferred if monomer sequence determinations are required.

Polyesters are encountered in many forms-from PET bottles, which are high-molecular-weight thermoplastics, to low-molecular-weight copolyesters used as coating resins for paint applications. Table 3a and b show the names and structure of glycols and acids commonly used commercially.

Polyethylene terephthalate (PET) is the most frequently encountered thermoplastic polyester and with slight modifications PET occurs as oriented films for packaging, reinforced PET for injection mouldings, fibres and bottles for carbonated drinks.

In the majority of these applications the PET is either unmodified or contains low levels of a second acid or glycol. The ^{1}H-NMR spectrum of PET at first sight is very simple; however, a more detailed examination yields more molecular structural information. We dissolve polymers of this type in perdeutero 1,1,2,2-tetrachloroethane (TCE) and obtain spectra at approximately 110 °C. To reduce degradation of the polymer whilst dissolving in TCE we prepare an amorphous film by quenching (see Sect. 2.1) from the melt. Figure 11 shows two vertical-scale plotted spectra of a commercial polymer used for bottles for carbonated drinks. The upper trace is scaled 5 times the lower trace. The resonances observed in the upper spectrum can be used to measure the following:

(i) the proportion of acids present as isophthalate; experience has shown that the resonance from H_c (see Fig. 11) yields the most reliable quantitative data;
(ii) the proportion of glycols present as diethylene glycol, typically in the range

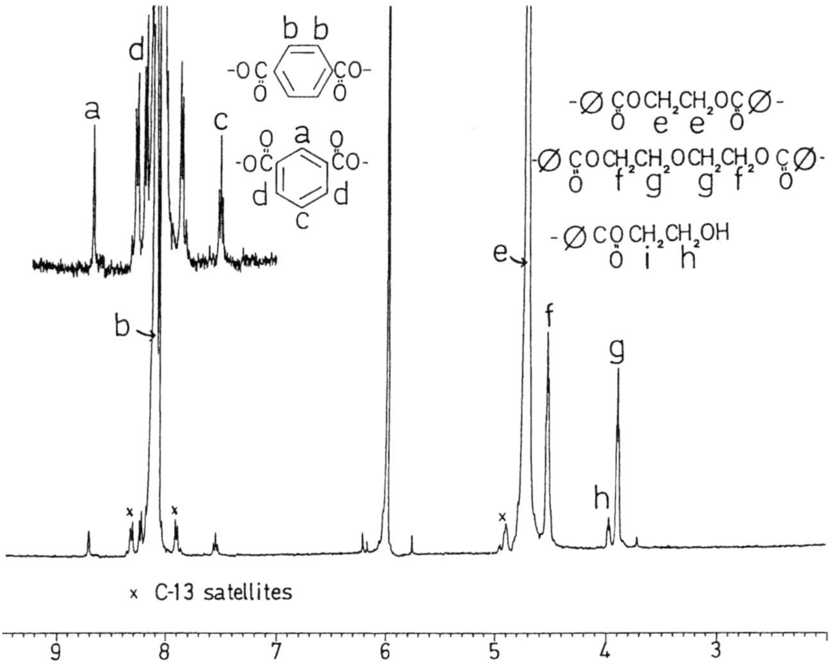

Fig. 11. 400 MHz ^{1}H-NMR spectrum of a polyethylene terephthalate which contains also copolymerised diethylene glycol & isophthalic acid. Upper spectrum X5 amplitude of lower one

1.5 to 4.0 molar percent;

(iii) an estimate of the $-\underset{\underset{\textstyle O}{\|}}{C}-O-CH_2CH_2-OH$ end group concentration.

For this last determination the polymer must be carefully dried before hot pressing so as to avoid thermal hydrolysis during preparation of the quenched film.

Modified polyethylene terephthalate polymers are used as coating resins onto PET for packaging applications. In most cases we find that the coating resins are soluble in deuterochloroform and a cold wash with this solvent will remove the coating resin. Figure 12 shows three spectra of this type and these examples were chosen in order to illustrate characteristic resonances. Isophthalate/terephthalate-based copolyesters are commonly encountered, for glycols ethylene and neopentyl glycols are most frequently used. The presence of an aliphatic dicarboxylic acid is detected by the resonance near 2.2-ppm from the $-CH_2-CH_2-\underset{\underset{\textstyle O}{\|}}{C}-$. In aromatic/aliphatic acid polyesters structure is observed on the resonance of the glycol methylene and this may be used for monomer sequence determinations (see Fig. 12c).

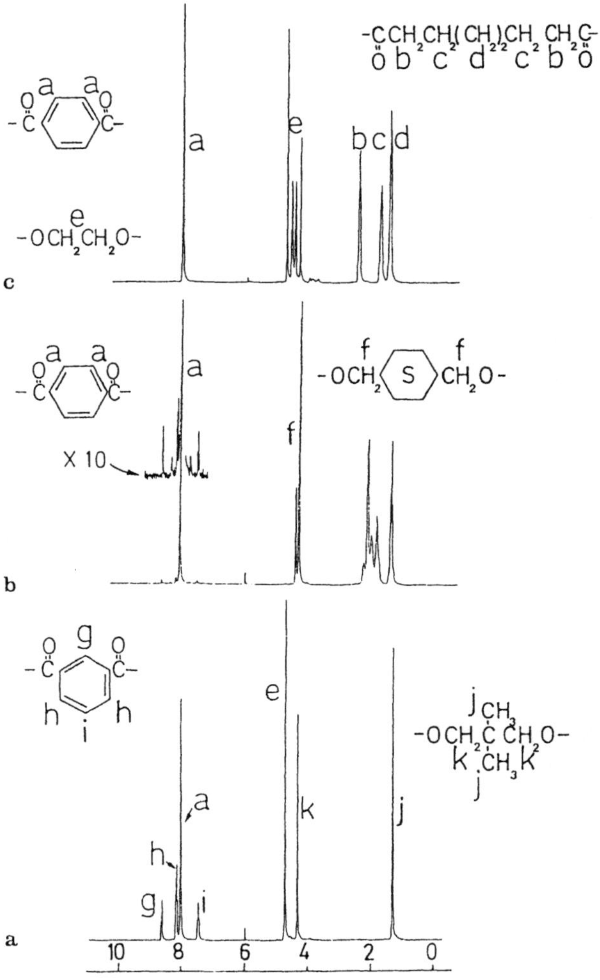

Fig. 12 a–c. 270 MHz ^{1}H-NMR spectra of 3 polyesters resins: (**a**) ethylene glycol/neopentyl glycol/isophthalate/terephthalate copolyester (**b**) cyclohexanedimethylol/terephthalate/isophthalate co-polyester (**c**) ethylene glycol/terephthalate/azelaic copolyester

Most copolyesters which are used commercially are 'random' in the sense that next-door neighbour sequencing yields a statistical distribution. In most cases ^{13}C-NMR spectra will ultimately yield more molecular structural information than ^{1}H-NMR. However, for most polyesters encountered ^{1}H-NMR yields sufficient data to identify and quantify the copolyesters. In many polymers sequencing information would not be necessary. In terephthalate copolyesters with two or more glycols ^{13}C-NMR more easily is able to yield sequencing information. Figure 13a shows a 67.9-MHz spectrum of a 60:40 mole EG/DEG/terephthalate copolyesters and the fine-structure on the *ipso* aromatic carbon confirms a 'random' distribution of EG/DEG units.

Two types of wholly aromatic copolyesters are commercially available and

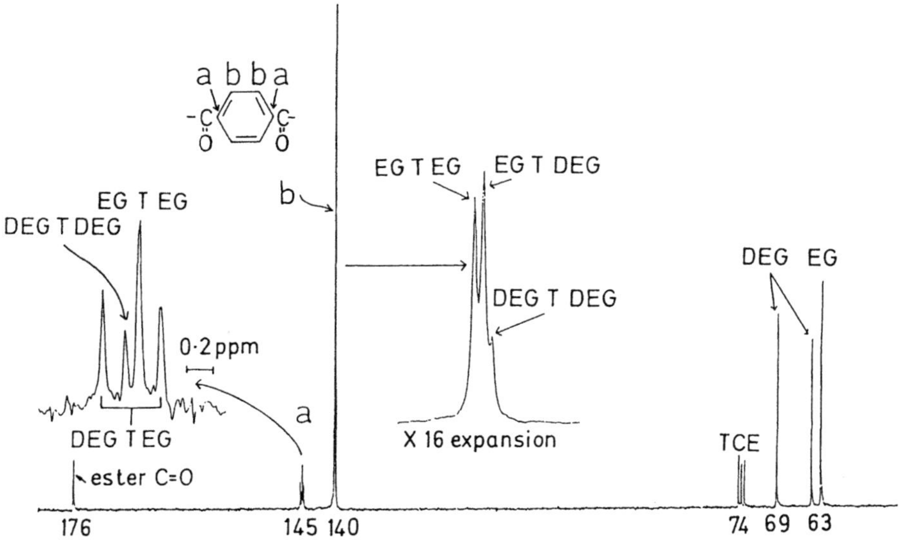

a

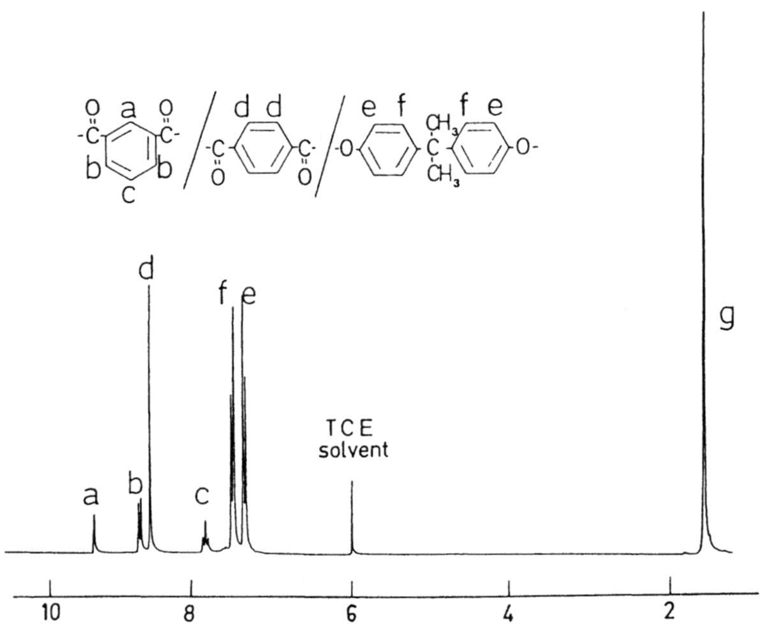

b

Fig. 13a. 67.8 MHz ^{13}C-NMR spectrum of an ethylene glycol/diethylene glycol/terephthalate copolyesters **(b)** 270 MHz ^{1}H-NMR spectrum of a polyarylate prepared from bisphenol A, terephthalate & isophthalate

these are:

(i) polyarylates, based on bisphenol A with terephthalate and isophthalate. These polymers dissolve in deuterochloroform and/or TCE and are fairly routine to analyse (Fig. 13b).

(ii) semi-crystalline polyesters based on hydroxybenzoate. This family of polymers known as either liquid crystal polymers (LCPs) or self-reinforcing polymers (SRPs) are not soluble in most solvents used for NMR studies and this presents difficulties in analysis.

Three approaches have been used successfully in our laboratory:

(i) dissolve the polymers at low concentration in a 70:30 v/v trifluoroacetic/deuteromethylene dichloride mixture;

(ii) obtain a CP-MAS solid-state spectrum;

(iii) hydrolyse the polymer in a 12:3:1 D_4 methanol/D_2O/NaOD mixture [66]. An ^{1}H-NMR spectrum of the hydrolysed mixture will characterise the monomeric species.

In these polymers the most frequently encountered monomeric units are:

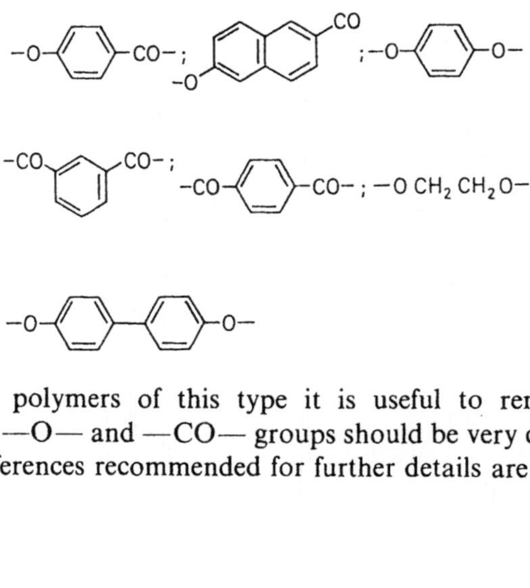

In analysing polymers of this type it is useful to remember that the stoichiometry of —O— and —CO— groups should be very close to unity.

Literature references recommended for further details are Refs. [67–71].

5 Polyamides

Commercially the most important polyamides are aliphatic amides known as nylons. They may be divided into two types: those prepared from condensation of a diamine and a dibasic acid, and those formed from the polymerisation of a lactam or amino acid. Homopolymers formed from the former route are known as nylon n_1/n_2 where n_1 and n_2 are the number of carbons in the diamine and diacid, respectively. Hence, nylon 6/10 is made from hexamethylene diamine and sebacic acid. Homopolymers prepared from the lactam route are referred to as nylon n_3, where n_3 is carbon number in the lactam, thus nylon 6 is the polymer prepared from caprolactam. Nylon copolymers can be prepared from several combinations of monomers; however, the most frequently encountered are nylon 6–6/6 copolymers.

Experience has shown that ^{13}C-NMR spectroscopy is a very powerful tool for identification, determination of composition and, for copolymers, monomer sequencing. ^{1}H-NMR is less widely used; however, we will give examples in which ^{1}H-NMR spectra are more useful.

^{13}C-NMR spectra of homopolyamides are usually straightforward and identification is relatively trivial. Figure 14 shows spectra taken in methane sulphonic acid of nylon 6, nylon 6/6 and nylon 11 polymers. Note that the spectra are simple and the numbers of resonances either equals or is half the number in the nylon type. Thus, nylon 6 and 11 spectra have 6 and 11 peaks respectively whereas nylon 6/6 has also 6. Thus, this simple approach does not

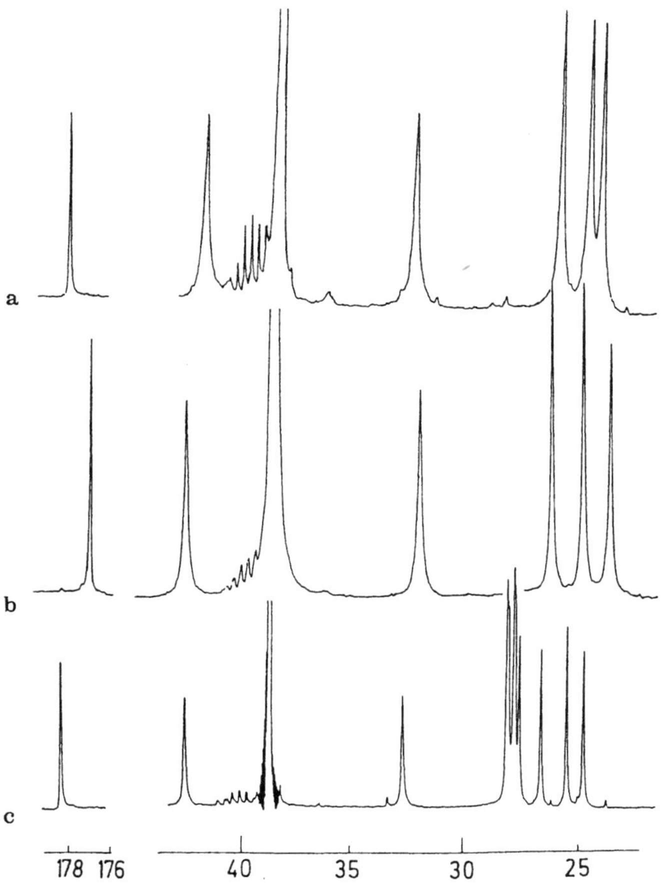

Fig. 14a–c. 67.8 MHz NMR spectra taken in methane sulphonic acid of polyamides: (**a**) nylon 6, (**b**) nylon 66, (**c**) nylon 11

distinguish between nylon 6 and nylon 6/6; however, these spectra combined with other data, e.g. reference spectra of known polymers or 'wet chemistry' hydrolysis and identification of diamine or diacids has led to ^{13}C-NMR being a very powerful tool for nylon characterisation.

More importantly the ^{13}C-NMR spectra of nylon copolymers can be used to:

(a) identify the comonomers
(b) quantify the concentration of each component and
(c) perform monomer sequence determinations in many polymers.

Nylon 6–6/6 copolymers are commercially very important and often are encountered as simple copolymers and as blends of copolymers with nylon 6 or 6/6 homopolymer. To achieve suitable data to distinguish between these possibilities, monomer sequencing data are required and this requires the use of strong acids as solvents. We find methane sulphonic acid suitable, Kricheldorf reports the use of fluoro sulphonic acid [72]. Spectra of nylon 6–6/6 copolymers in formic acid do not show sequencing information. Figure 15 shows the carbonyl resonances of a nylon 6–6/6 copolymer, the four resonances arise from the sequences:

$$\text{—NH(CH}_2\text{)}_6\text{NH*}\underset{\overset{\|}{\text{O}}}{\text{C}}\text{(CH}_2\text{)}_4\underset{\overset{\|}{\text{O}}}{\text{C}}\text{NH(CH}_2\text{)}_6\text{NH—} \qquad 6/6\text{–}6 \quad \text{'a'*}$$

$$\text{—NH(CH}_2\text{)}_6\text{NH—}\underset{\overset{\|}{\text{O}}}{\text{C}}\text{(CH}_2\text{)}_4\text{*}\underset{\overset{\|}{\text{O}}}{\text{C}}\text{NH(CH}_2\text{)}_5\underset{\overset{\|}{\text{O}}}{\text{C}}\text{—} \qquad 6/6\text{–}6 \quad \text{'b'*}$$

$$\text{—HN(CH}_2\text{)}_5\text{*}\underset{\overset{\|}{\text{O}}}{\text{C}}\text{NH(CH}_2\text{)}_6\text{NH—} \qquad 6\text{–}6/6 \quad \text{'c'*}$$

$$\text{—NH(CH}_2\text{)}_5\text{*}\underset{\overset{\|}{\text{O}}}{\text{C}}\text{NH(CH}_2\text{)}_5\underset{\overset{\|}{\text{O}}}{\text{C}}\text{—} \qquad 6\text{–}6 \quad \text{'d'*}$$

*in Fig. 15.

On this polymer the proportions of the four sequences are consistent with the polymer having a random distribution of monomeric units.

Occasionally terpolymers are encountered and one such example is a 6–6/6–6/10 terpolymer (spectrum taken in CH_3SO_3H is given in Fig. 16). The resonances from the methylene groups confirm the polymer is composed of 6, 6/6 and 6/10 nylon segments. The complexity of the resonances from the carbonyl group can be used to show this polymer is a terpolymer. In Fig. 16 nylon 6 resonances are indicated by *x*'s; 66 by *o*'s and 610 by +'s.

Although in this article we have only been able to give a few examples of identification and characterisation of nylon polymers by ^{13}C-NMR, many more exist and for further examples the reader is referred to the series of excellent papers by Kricheldorf [73–82].

^{1}H-NMR spectra of nylon polymers are generally less useful than ^{13}C-NMR spectra. However, some nylon polymers can be analysed by ^{1}H-NMR, such as

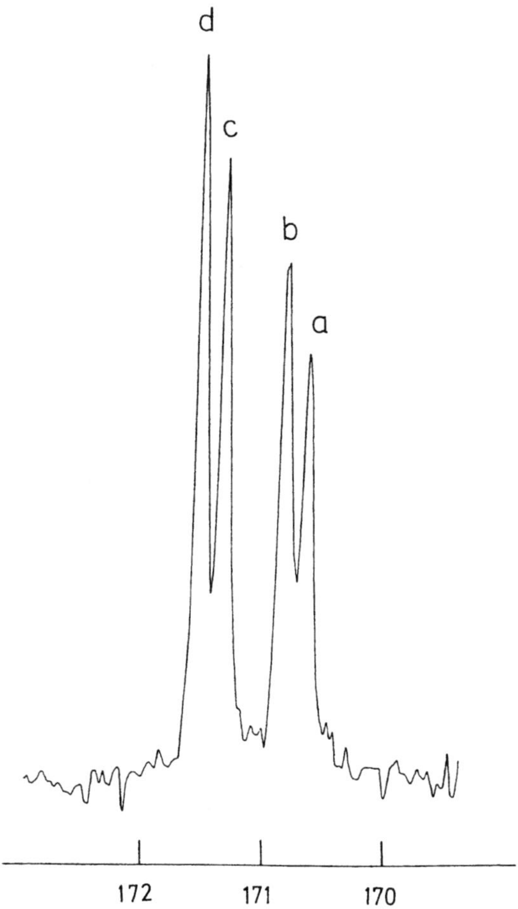

Fig. 15. The resonances from the amide carbonyl group carbon in the ^{13}C-NMR spectrum of a nylon 6/66 random copolymer

mixed aromatic/aliphatic nylons, nylon fibres which contain low levels of aromatic monomers, or polyethers such as polyethylene glycol or polypropylene glycol. For ^{1}H-NMR studies we use hexafluoro isopropanol (HFIP) as solvent. This is a non-deuterated solvent and for magnet homogeneity adjustment two methods have been used successfully:

(i) shimming on the free induction decay (FID) or
(ii) the HFIP solution is contained in a 5-mm tube and this is placed concentrically inside a 10-mm tube which contains deuterochloroform; the deuterochloroform is to provide a deuterium resonance for field homogeneity adjustment.

6 Polyarylene Ether Sulphones and Arylene Ether Ketones

Arylene ether sulphone, arylene ether ketone homopolymers and copolymers are high-temperature thermoplastics which have been available for several years. The most familiar of this class of polymer are amorphous linear polysulphones

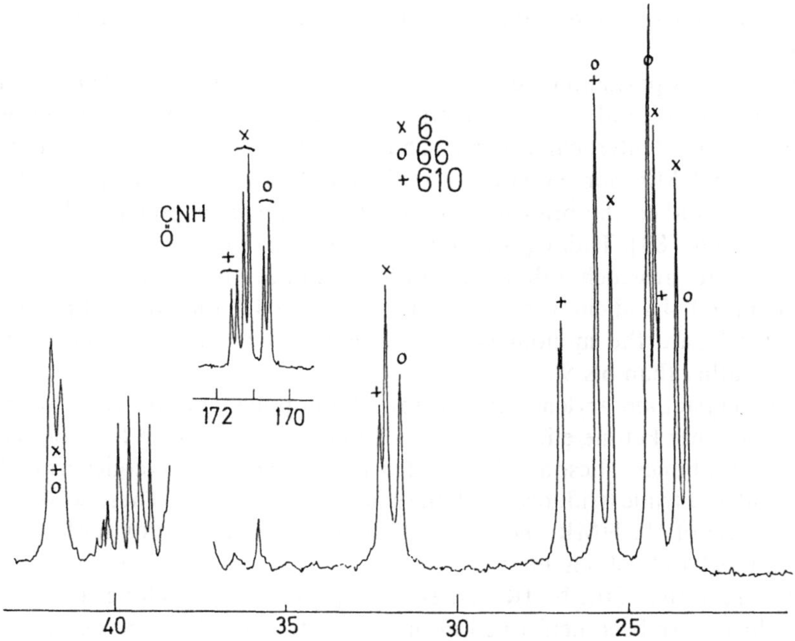

Fig. 16. 67.8 MHz ^{13}C-NMR spectrum of nylon 6/66/610 terpolymer taken in methane sulphonic acid. Resonances from 6 indicated by *x*, 66 by ○ and 610 by +

such as "Victrex" PES polysulphone (I), UDEL (II) and the semi-crystalline polymers such as polyether ketones (III) (PEK), polyetheretherketone (IV), "Victrex" PEEK, the recently introduced poly ketone (V) (PEKEKK).

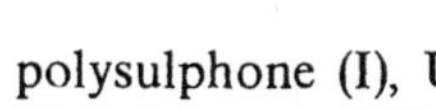

I

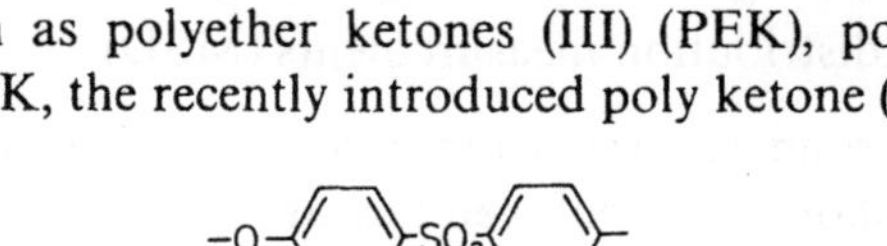

II

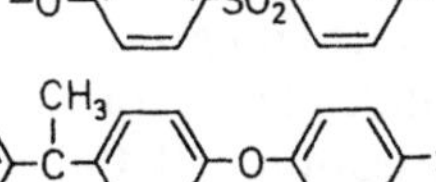

III

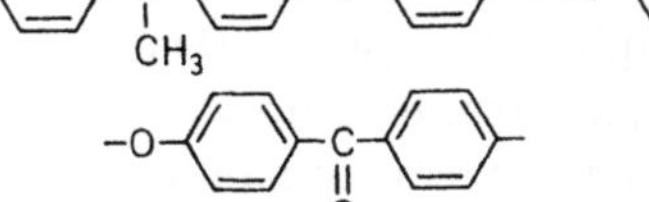

IV

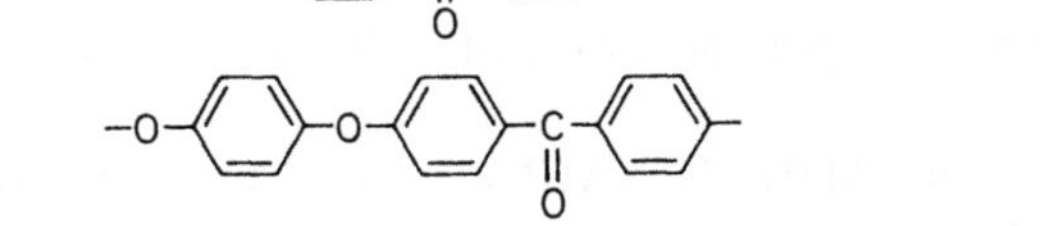

V

In recent years polymers which are a variation of the ether sulphone-ether ketone theme have appeared. In general, all these polymers have regular all-*para* repeating units and, as a consequence, have very simple ^{13}C-NMR spectra. Copolymers in which the ether and sulphone and/or ketone groups are not regularly distributed often have advantageous thermal and mechanical behaviours.

The amorphous polysulphones are soluble in solvents such as a deuterated dimethyl sulphoxide and 1,1,2,2-tetrachloroethane. The UDEL polymer is readily soluble in deuterochloroform. Thus, polymers of this type can be evaluated by ^{1}H-NMR. Some very early work on polysulphone was reported by our group and studied the branching levels of PES polymer prepared by the electrophilic route [83]. Today polysulphones are prepared commercially by a nucleophilic route in which a dihalide and bisphenol are polycondensed [84]. Thus, to prepare I, 4,4′-dichlorodiphenyl sulphone and 4,4′-dihydroxy diphenyl sulphone (bis *S*) are the monomers used, whereas in the preparation of II, bisphenol *A* rather than bis *S* is used.

The semi-crystalline arylene ether ketone homopolymers and to a lesser extent arylene ether ketone/ether sulphones copolymers are much less soluble than the polysulphones. These are dissolved in strong acids such as concentrated sulphuric acid and triflic (trifluoro methane sulphonic) acid [85]. These solvents preclude the use of ^{1}H-NMR and only ^{13}C-NMR is used to study molecular structure of this family of polymers.

The great majority of the NMR studies of the polysulphone/ketone family of polymers which have been performed in our laboratory has been on polymers in which 1,4-phenylene groups ϕ links have been joined by ether, sulphone and ketone functions (denoted by E, S & K). When two ϕ groups are joined directly, then of course we have a diphenyl unit and this unit is denoted by *D*.

The NMR studies on polymers of this type has fallen into two categories:

(i) end-group analysis;
(ii) monomer sequencing distribution measurements (MSD).

The importance of end-group analysis cannot be overstated, as the thermal performances of polysulphone and polyketones are governed by having the correct end group. To achieve the correct end groups, the polymers are usually prepared with a slight excess of dihalide. Thus, for instance, Victrex PES was prepared by the nucleophilic polymerisation of 4,4′-dichlorodiphenyl sulphone (DCDPS) and 4,4′-dihydroxydiphenyl sulphone (bis *S*) for 20 h at 300 °C in diphenyl sulphone, with potassium carbonate present in an amount of just over 1 mol per mol of total bis *S*. A small excess of halide was added to control the molecular weight. All the copolymers were of molecular weight (M_n) greater than 8000.

Figure 17 shows a 400 MHz ^{1}H-NMR spectra of a typical PES polymer and the assignments of the resonances are shown in this figure. Many applications of *E*, *S*, *K* polymers require functionalised end groups [86] and the ability of ^{1}H-NMR to relatively quickly determine the type and proportion of end groups in polysulphones in particular has assisted greatly polymer chemists.

^{13}C-NMR has been shown to be a very powerful tool for sulphone/ketone copolymers [87–90]. To illustrate the power of this technique we will consider two systems:

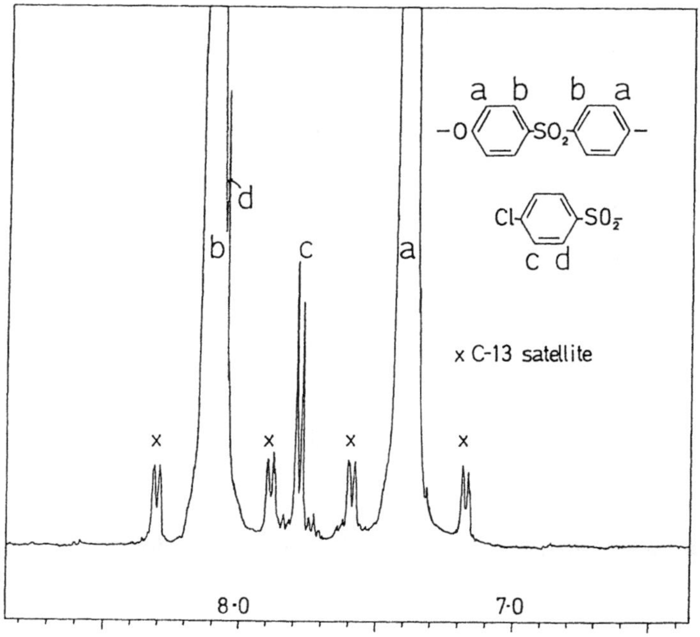

Fig. 17. 400 MHz ^{1}H-NMR spectrum of polyarylene ether sulphone in D_6-DMSO solution

(a) Type A, a 50:50 copolymer of $-O\phi\ SO_2\phi-/O\phi SO_2\phi\phi SO_2\phi$ (ES/ESDS) (soluble d_6-DMSO);
(b) Type B, a 50:50 copolymer of $-O\phi CO\phi$-/-$O\phi CO\phi CO\phi$-(EK/EKK)-semi-crystalline and examined in concentrated sulphuric acid.

The need to know the MSD of these copolymers arises because the mechanical and thermal behaviour are determined by the molecular structure. Also, the MSD determines whether a polymer is semi-crystalline or amorphous. To appreciate how the MSD of these polymers can vary the concept of transetherification needs to be introduced. One type of transetherification for ether interchange occurs in nucleophilic polymerisation when ether linkages are cleaved by phenoxide end groups [91]. The ether bond between benzene nuclei is usually stable to nucleophilic attack, but the ether linkages in, for instance, poly(aryl ether sulphone) are activated by the sulphone groups so that fission of the ether linkage is possible by, for instance, phenoxide end groups, and transetherification occurs.

The copolymer of type A, experimental polymer KM52, has a Tg of 260 °C which is 30 °C higher than the Tg of PES. In the absence of transetherification during polycondensation, the 50:50 copolymer of this type should have a regular all-*para* repeating unit as shown in **I**:

$$—O\phi SO_2\phi O\phi SO_2\phi\phi SO_2\phi— \qquad \textbf{I}$$

(denoted by ESESDS)

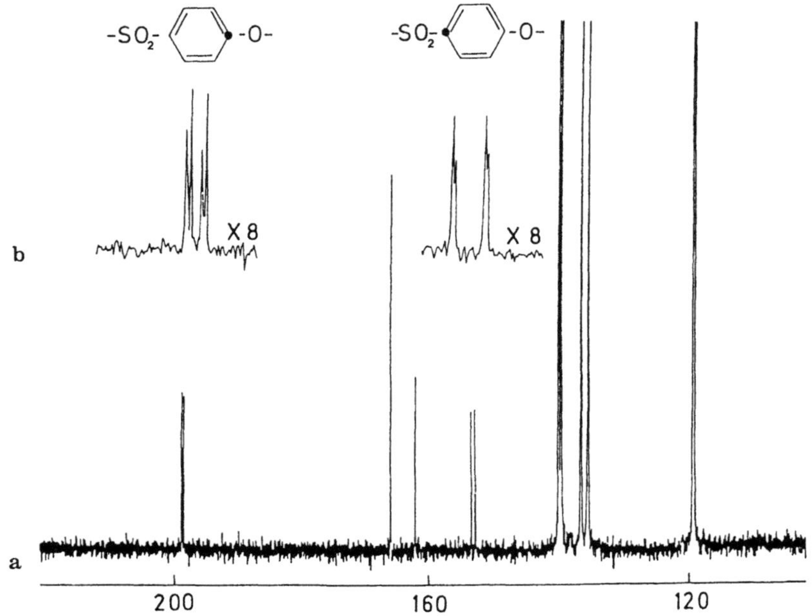

Fig. 18a. 100 MHz ^{13}C-NMR spectrum of a 50/50-$\phi SO_2\phi O$-/-$\phi SO_2\phi\phi SO_2\phi O$ copolymer. (**b**) expansion of the resonances from the ipso carbons adjacent to ether and sulphone groups

Transetherification during polycondensation would eventually lead to a random copolymer of units of ES and ESDS.

Figure 18a shows a ^{13}C-NMR spectrum of a copolymer of this type. The assignments given on this figure were made from empirically calculated chemical shifts and by comparison with spectra of known polymers, e.g. PES. Expansion of this spectrum shows that the resonances near δ 136-ppm and δ 159-ppm each consists of two pairs of equally intense doublets (Fig. 18b). If structure **I** was the repeating unit of this polymer, then two singlets would be expected in each of these regions. The effect of transetherification at the ether groups is to scramble the units in the polymer and so to change the sequence distribution. Thus, on transetherification, possible sequences with a common central ES unit are:

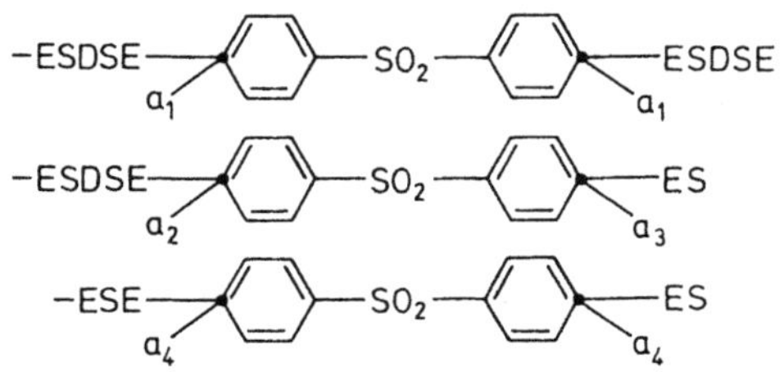

Four different environments, a_1–a_4, are possible for the carbons denoted by the filled circles. Carbon type a_4 is found in PES and by observation of the spectrum of a mixture of PES and the copolymer whose spectrum is shown in Fig. 18a the δ 159.09-ppm resonance was identified as due to carbon type a_4. Reference to the spectrum of polymer with the repeating unit:

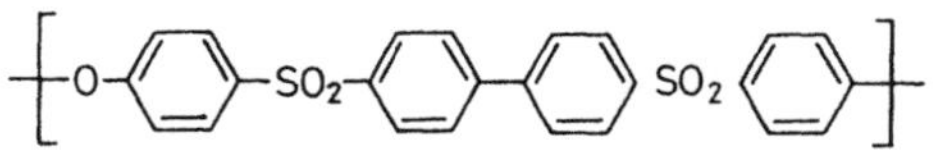

and the changes in the spectra of these high-Tg copolymers when the ratio of ES and ESDS units is changed, led to the conclusion that the doublets were caused by the effects of neighbouring units. In the case of 'a'-type carbons, it is concluded that $a_1 = a_3$ and $a_2 = a_4$. This confirms that this polymer is a 50:50 random copolymer of ES and ESDS units.

The ^{13}C-NMR chemical shifts for some of the carbon nuclei in the polymers of this type are sensitive to functional groups situated three or four aromatic nuclei along the polymer chain. In polyether ketone polymers even longer range effects are observed and this is illustrated in Fig. 19 which shows two spectra of 50:50 EK/EKK copolymers. Polyketone A has a regular structure: (EKEKK)n; whereas we were able to demonstrate [90] that polyketone B was essentially a random copolymer of EK and EKK units.

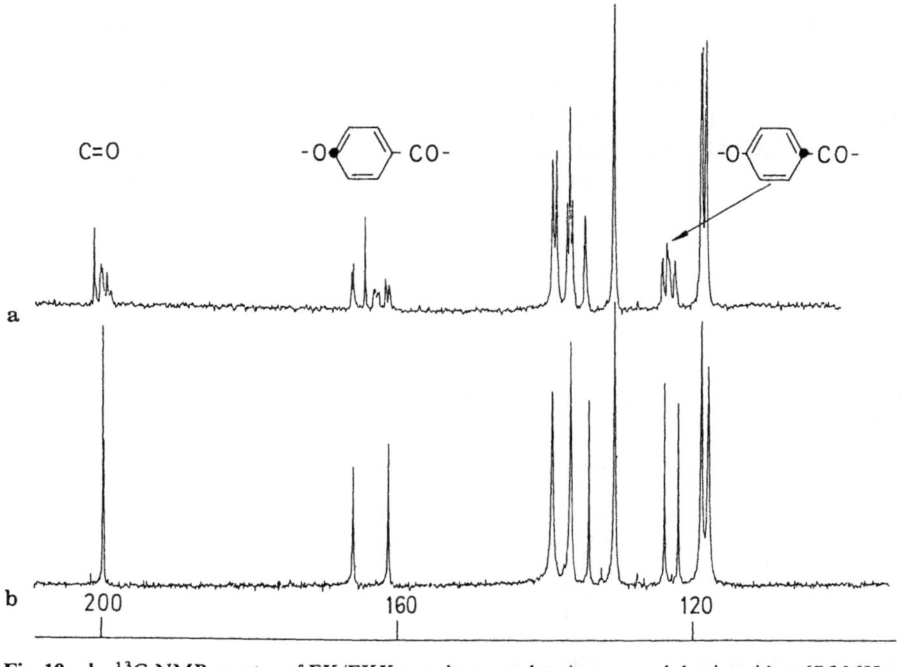

Fig. 19 a, b. ^{13}C-NMR spectra of EK/EKK copolymers taken in conc. sulphuric acid at 67.8 MHz. (**a**) an essentially random copolymer of EK and EKK units (**b**) alternatively EK/EKK copolymer

On polymers of this type discussed in this section, ^{13}C-NMR provides data on molecular structure which cannot be derived by any other technique. Work in this area continues at a high level as many companies proceed towards the high-temperature, high-performance goal.

7 Polyethers

Polyethers are commercially available as thermoplastics, primarily oxymethylenes, and a wide range of homo- and copolymers based on ethylene oxide and propylene oxide. ^{1}H-NMR spectroscopy provides a satisfactory method for the qualitative and quantitative analysis of these polymers. The chemical shift values for typical ^{1}H-NMR spectra of polyethers are given in Table 4.

Figure 20 shows a 270 MHz ^{1}H-NMR spectrum of an oxymethylene/oxyethylene copolymer taken in a D_2-TCE at 100 °C. The resonance from the low level of incorporated oxyethylene arise at 3.7-ppm and the resonance for oxymethylene adjacent to the oxyethylene at 4.8-ppm. This shows that in this polymer the oxyethylene units are almost, if not entirely, all present as isolated units.

The polyethers based on propylene oxide (PO) and/or ethylene oxide (EO) can be identified by ^{1}H-NMR and composition of these polymers can be determined from the integrals of the resonance of the methyl group (from PO) and the total resonances from the methine/methylene group near 3.6 ppm.

Monomer sequencing and end-group analysis of EO/PO polymers is best achieved by ^{13}C-NMR. Figure 21 shows ^{13}C-NMR spectra of (a) polypropylene oxide (b) polyethylene oxide (c) a random EO/PO copolymer, taken in $CDCl_3$ at 100 MHz ^{13}C-NMR frequency. The end-group resonances are from $-CH_2-CH_2-OH$ from EO end groups and $-CH_2-\underset{\displaystyle CH_3}{\underset{|}{CH}}-OH$ from PO end groups.

Table 4. Resonances of protons in polyalkylene oxide and alkylene oxide copolymers

Backbone	Group	Chemical shift value ppm
$-OCH_2O-$	$-CH_2-$	4.9
$-OCH_2CH_2O-$	$-CH_2-$	3.5
$-OCHCH_2O-$	$-CH-$	3.7
|	$-CH_2-$	3.6
CH_3	$-CH_3-$	1.2
$OCH_2CH_2CH_2CH_2O$	$-CH_2-(1)$	3.2
1 2 2 1	$-CH_2-(2)$	1.6

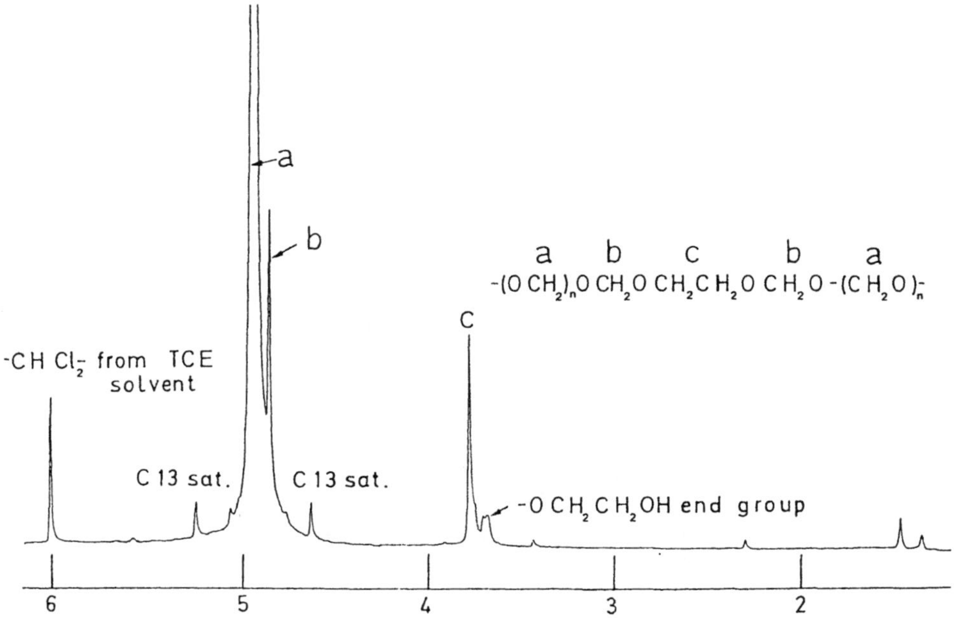

Fig. 20. 270 MHz ^{1}H-NMR spectrum of oxymethylene/oxyethylene copolymer taken in D_2-1,1,2,2-tetrachloroethane at 100 °C

The multiplicity of the PO end-group methine resonances in Fig. 21a is a result of the tacticity of the essentially atactic PPO.

The ^{13}C-NMR spectra EO/PO copolymers are complex (see Fig. 21c). However, the complicated ^{13}C-NMR spectra are caused by detection of many carbons in slightly different environments. Heatley and co-workers [27] have made a very detailed analysis of the ^{13}C-NMR spectra of EO/PO copolymers and the assignment of the resonances observed in Fig. 21c are taken from Heatley's work [92]. The nomenclature used is: propylene oxide ≡ P where the arrow runs from CH_2 tail to $CH(CH_3)$ head, ethylene oxide ≡ *E*; end groups are denoted by *E** and *P**. Where an unsymmetrical unit exists, methylenes from ethylene oxide are represented by *a* or *b* for left- or right-hand methylenes. In a propylene oxide unit, *t*, *s* and *p* distinguish methine, methylene and methyl carbons, meso diads are designated by *m*, racemic diads by *r*.

EO/PO copolymers are often encountered as multiblock polymers and several different initiators are used commercially. The ^{13}C-NMR spectrum of EO/PO copolymers taken under quantitative conditions can yield:

(i) composition;
(ii) type and concentration of end groups as triads, i.e. PPP*, EPP*, EEP*, EEE*, PEE* and PPE*
(iii) diad and triad MSD sequences;

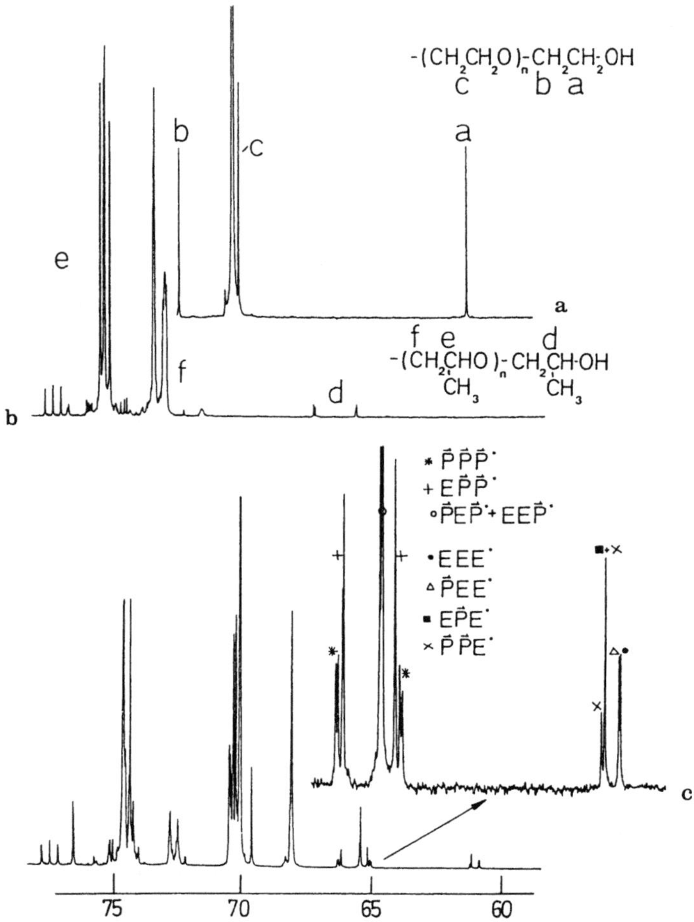

Fig. 21 a–c. 100 MHz ^{13}C-NMR spectra of polyethers. (**a**) polyethylene oxide, (**b**) polypropylene oxide, (**c**) a random EO/PO copolymer

(iv) identification of initiator type and whether first unit after initiator is EO or PO.

The combined data from (i) to (iv) can in many cases be used to determine an average polymer structure.

8 Miscellaneous Vinyl Polymers

Miscellaneous vinyl polymers cover a very wide range of polymer types. In order to limit the length of this section, and yet give the impression of the power of NMR to evaluate this type of polymer, the author has selected to discuss only polymers based on vinyl chloride and styrene. In each case the homopolymers and selected copolymers will be used as illustrations. Literature references to further NMR studies of other copolymers will be given where appropriate.

8.1 Vinyl Chloride Polymers

Measurements of the tacticity of polyvinyl chloride has been widely reported over many years. ^{13}C-NMR spectroscopy provides a suitable method for this measurement and pentad sequences can be determined from the carbon resonance of the methine group. The 2D-NMR approach to confirmation of the assignments of the ^{13}C-NMR resonances by Levy and co-workers [23] seems to have once and for all provided definitive assignments. These are given in the spectrum of PVC in Fig. 22.

PVC polymerised at lower temperature is more syndiotactic than polymer prepared at elevated temperatures. Typical triad percentages for a range of polymers are shown in Table 5. PVC is commercially very important and for this reason much NMR has been done on determination of branching in this polymer. Elegant work by Starnes and co-workers has shown that the majority of branches are:

$>CH—CH_2—Cl$

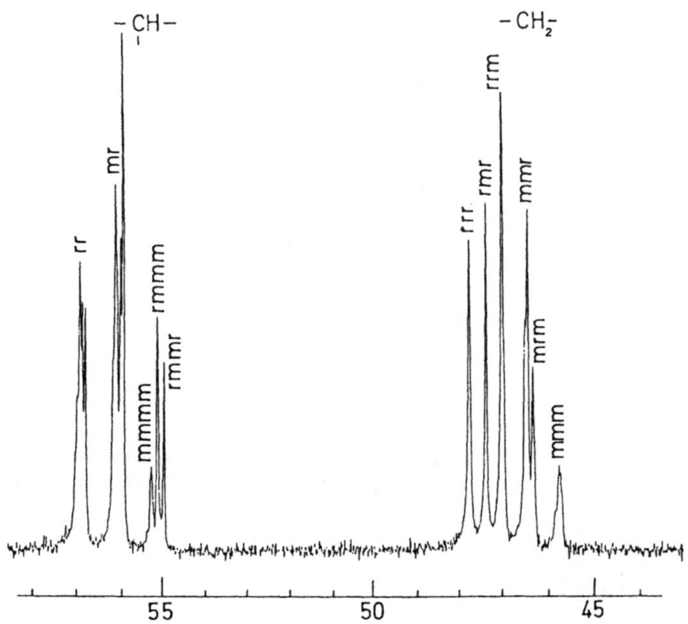

Fig. 22. 67.8 MHz ^{13}C-NMR spectrum of polyvinyl chloride taken in D_2-TCE

Table 5. Triad percentages for PVCs prepared in the temperature range −50 °C to +70 °C

Polymerisation temperature (°C)	% mm triad	% mr triad	% rr triad
70	22	48	30
50	20	48	32
35	18	50	32
−10	14	50	36
−30	10	50	40
−50	7	50	42

with other branches at lower levels: $\rangle CClCH_2CH_2Cl$ and $\rangle CClCH_2CHClCH_2CH_2Cl$ [93–98].

Acrylic polymers are often added to PVC at low levels as processing aids and the ^{1}H-NMR spectrum of a polymer of this type is shown in Fig. 23. This type of spectrum can be achieved in a few minutes on current commercial spectrometers and the level and type of processing aid can be identified without too much difficulty.

Vinyl chloride/vinyl acetate copolymers are available commercially with a wide range of compositions. ^{1}H-NMR spectroscopy provides a satisfactory method for determination of composition, and Fig. 24 shows 400 MHz ^{1}H-NMR

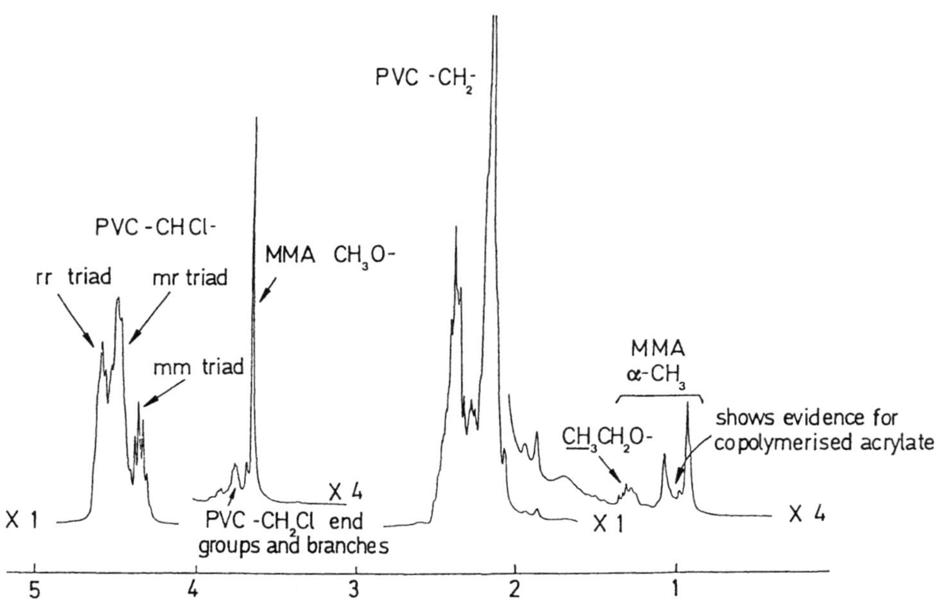

Fig. 23. ^{1}H-NMR spectrum of polyvinyl chloride which contains ca. 2 wt% of a methyl methacrylate/ethyl acrylate copolymer processing aid

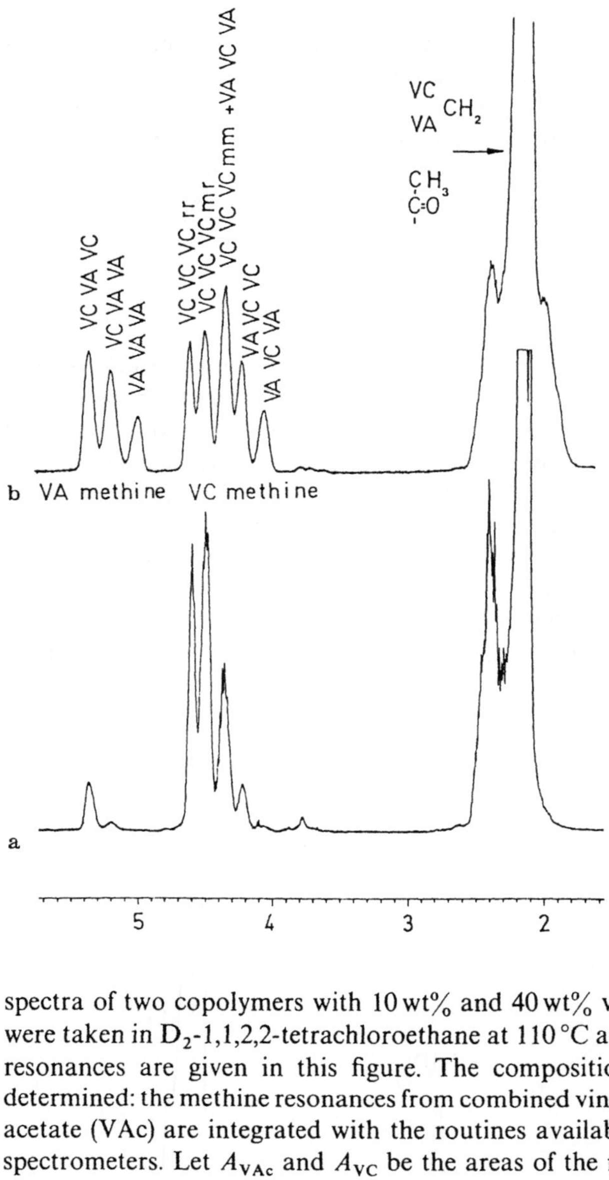

Fig. 24a, b. 400 MHz ^{1}H-NMR spectra of vinyl chloride/vinyl acetate copolymers (**a**) ca. 10 wt% VAc, (**b**) ca. 40% VAc

spectra of two copolymers with 10 wt% and 40 wt% vinyl acetate. The spectra were taken in D_2-1,1,2,2-tetrachloroethane at 110 °C and the assignments of the resonances are given in this figure. The composition of the copolymers is determined: the methine resonances from combined vinyl chloride (VC) and vinyl acetate (VAc) are integrated with the routines available on commercial NMR spectrometers. Let A_{VAc} and A_{VC} be the areas of the methine resonances from VAc and VC, respectively.

Then the mol ratio $\frac{VAc}{VC} = \frac{A_{VAc}}{A_{VC}}$

and the wt ratio $\frac{A_{VAc}}{A_{VC}} \cdot \frac{86}{62.5} = x$

then wt% VAc $= \frac{x}{1+x} \cdot 100$

In our laboratory we have established a plant quality control test based on infrared spectroscopy. In this method thick plaques (4 → 18 mm in thickness depending upon VAc level) were used to determine the absorbance/cm of the 4675 cm^{-1} near infrared band. A very good correlation was obtained between these IR data and the wt% VAc by NMR which indicated that for copolymers of ≦40 wt% VAc the absorbance of the 4675 cm^{-1} band is not affected by the environment of the VAc, i.e. whether VC VAc VC, VC VAc VAc, or VAc VAc VAc.

8.2 Styrene Polymers

Polystyrene was the subject of a very early paper by Bovey and co-workers [99]. In general, the ^{1}H-NMR spectra of polystyrene and copolymers are broader than normal encountered for other polymers. The ^{13}C-NMR spectra of polystyrene prepared by free radical polymerisation show nine distinct methylene resonances but only a single broad methine resonance. Many papers have appeared which discuss the assignment of these peaks and similar numbers of peaks from the quarternary carbon. However, from an industrial viewpoint these measurements are probably not too important.

Although spectra from styrene copolymers are relatively broad, they are very useful for qualitative and quantitative analysis. This is illustrated in Fig. 25 which

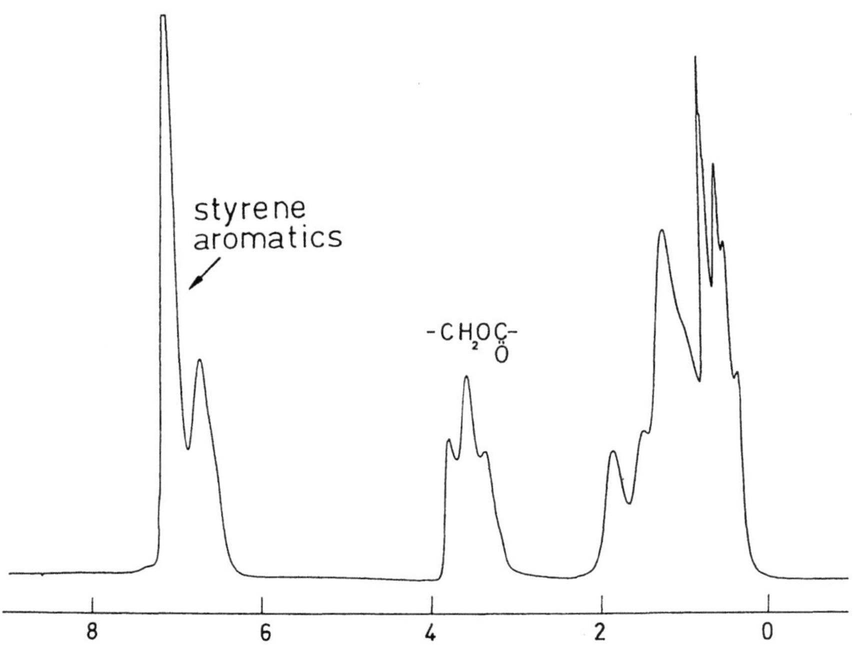

Fig. 25. ^{1}H-NMR spectrum of a styrene/ethyl acrylate (44 : 56 w/w) copolymer taken in $CDCl_3$ at 270 MHz

is a 270 MHz ^{1}H-NMR spectrum of a styrene/ethyl acrylate copolymer. The composition of the copolymer can be determined straightforwardly from the styrene aromatic protons near 7-ppm and ethyl acrylate —CH_2—O—CO— resonance around 3.5-ppm.

As for polystyrene, spectra of styrene/methyl methacrylate copolymers have been very fruitful for literature work [100–105]. However, the major need for the author required for spectra of styrene copolymers has been in the identification of multistage acrylic polymers where one phase is often a styrene/acrylate or methacrylate copolymer. In analysis of these polymers, ^{13}C-NMR spectra are of most use and reference spectra of known copolymers are invaluable. As well as in-house reference spectra, literature data are very useful and references used include the following:

styrene/methyl methacrylate (Ref. [105]);
S/methyl acrylate (Ref. [106]);
S/several acrylates (Ref. [107]);
S/*n*-butyl acrylate (Ref. [108]).

9 ^{19}F-NMR

^{19}F-NMR has been used to (a) study end groups in polymers (b) determine the molecular structure of fluoropolymers. The latter being the major application.

The study of fluorocarbon resins by ^{19}F-NMR spectroscopy reveals some interesting structural features which do not occur in the corresponding chlorocarbon resins. Thus the ^{1}H-NMR spectrum of polyvinylidene chloride has only one resonance (at about 4.9-ppm) from which it is deduced that all protons in the polymer chain are equivalent and that the polymerisation is head-to-tail (H-to-T) and the polymer repeat is —CH_2—CCl_2—.

By analogy, the ^{19}F-NMR spectrum of polyvinylidene fluoride (PVdF) would consist of one resonance. However, the ^{19}F-NMR spectrum of PVdf shows four resonances. From model compound comparison these resonances were identified with CF_2 groups marked CF_2^* in the following sequences [109].

—$CH_2CF_2CH_2CF_2^*CH_2CF_2$—

—$CF_2CH_2CH_2CF_2^*CH_2CF_2$—

—$CH_2CF_2CH_2CF_2^*CF_2CH_2$—

—$CF_2CH_2CF_2CF_2^*CH_2CH_2$—

Thus these data show that PVdF has an irregular structure because some H-to-H polymerisation occurs.

Head-to-head polymerisation has been shown by NMR to occur in other soluble fluoropolymers: VdF/hexafluoropropylene copolymers [110]; polyvinyl fluoride [111]; trifluoroethylene/isobutene copolymers [112].

This surprisingly detailed analysis of fluorocarbon resins from their ^{19}F-NMR spectra is possibe because chemical shifts between ^{19}F nuclei in different environments are very large compared with the chemical shifts observed in ^{1}H-NMR spectra. This means also, provided suitable solvents can be found and the resonances can be assigned, that quantitative analysis of fluorocarbons is comparatively straightforward. For further information on ^{19}F-NMR spectra of fluorocarbons the reader is referred to two reviews on this subject [113, 114].

Fluorine end groups are present in some poly ether ketones e.g. PEEK and Devaux and coworkers [115] have shown that the quantitative measurements of the signals obtained can be used to determine number average molecular weights (M_n). The M_n values determined by ^{19}F-NMR agreed well with results from gel permeation chromatography.

Derivatisation of hydroxy end groups in polyethers by trifluoro-acetic acid has been reported [116, 117]; and hydroxyls in prepolymers by hexafluoro-acetone [118].

10 Polymer Blends

Polymer blends have been used for several years to tailor physical properties to specific requirements. The addition of a rubbery component to a rigid polymer is well known to improve toughness. Examples include ethylene/vinyl acetate in polyvinyl chloride and polyolefin rubber in polypropylene. NMR has been used for several years in the qualitative analysis of blends of these types. The vast majority of NMR work reported in the literature on polymer blends uses solid-state NMR to investigate compatibility.

In this report we shall consider examples where chemical reactions occur when two polymers are blended thermally. Transesterification occurs in many polyester-polyester blends when they are melt extruded. In other polymer blends chemical reaction may be proposed between functional groups in the different polymers. NMR can often be used successfully to study interactions of this type and this leads to potential:

(i) improvement in the process;
(ii) quality control tests, and
(iii) patent specification for preferred products

In this section we will consider a few examples where NMR has confirmed that chemical reactions have occurred when blends of polymers were processed.

Blends of nylon homopolymers will amide-exchange at processing temperatures and form block copolymers. In our laboratory we have demonstrated by ^{13}C-NMR spectroscopy that nylon 6 and nylon 66 homopolymer blend amide exchanged at 285 °C. At this temperature for 1 hour 10% of the polymer dyads were 6/66 interchange units.

Mixtures of polyesters when extruded can transesterify and we have shown by ^{1}H-NMR spectroscopy that a polyethylene terephthalate (PET)/polyethylene naphthoate mixture transesterified to produce the interchange.

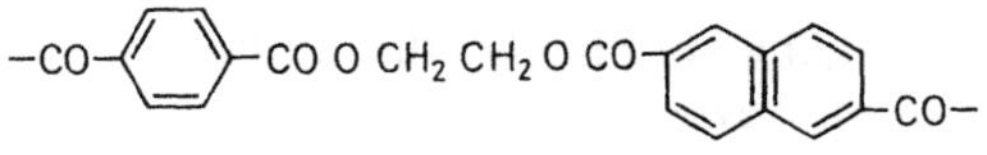

For PET blends with a terephthalate/isophthalate/bisphenol A polyarylate ^{1}H NMR was used to measure transesterification. The resonance from the hydrogen flanked by the two carbonyl groups in the isophthalate was most sensitive to environment and three resonances were observed on highly transesterified blends. These originate from the sequences:

(i) bis A – IPA – bis A
(ii) bis A – IPA – EG, and
(iii) EG – IPA – EG.

The former is present in the starting polyarylate and the latter in isophthalate-modified PETs (see Fig. 11).

Pillon and Ultracki [119] have reported on the preparation, and confirmed the molecular structures from NMR spectra, of a nylon 66/polyethylene terephthalate (PET) block copolymer prepared to act as a compatibiliser between nylon 66 and PET.

Devaux and co-workers have reported on a very thorough study of transesterification of bisphenol A polycarbonate with polybutylene terephthalate [120–122] and polyethylene terephthalate [123/124]. In both studies ^{1}H- and ^{13}C-NMR data are used to measure the extent of transesterification and deduce reaction mechanisms.

11 Conclusion

In this article the author aimed to give an impression of the very important role that solution-state NMR has in an industrial polymer laboratory. As is usual in this environment for reasons of confidentiality, several potential examples are not publishable; however, the author hopes that those given will have created a favourable impression of the power of NMR in this area.

As stated in the body of the article, solution-state NMR is often used to solve problems without assistance from other techniques. However, the more important difficult-to-solve problems encountered always benefit from a multi-discipline approach.

Acknowledgement: The author would like to acknowledge his many colleagues in ICI who have assisted in the studies reported here. In particular, the staff of the

NMR team at Wilton and the numerous polymer chemists who have prepared the polymers used for these studies. The importance of both these groups cannot be overstated, especially the polymer chemists who have prepared model polymers and reference materials which enable us, the NMR spectroscopists, to confirm proposed assignment. Finally, the author is indebted ICI plc for permission to publish this review.

12 References

1. Fisch MH, Dannenberg JJ (1977) Analytical Chemistry 49: 1405
2. Paxson JR, Randall JC (1978) Anal. Chem. 50: 1777
3. Wehrli FW, Merchand AP, Wehrli S (1988) Interpretation of carbon-13 spectra, Wiley, New York
4. Derome AE (1987) Modern NMR techniques for chemistry research, Pergamon Press
5. Abraham RJ, Fisher J, Loftus P (1988) Introduction to nuclear magnetic resonance spectroscopy, Wiley, New York
6. Randall JC (1977) Polymer sequence determinations-carbon-13 NMR method, Academic Press, New York
7. Bovey FA (1972) High resolution NMR of macromolecules, Academic Press, New York
8. Bovey FA (1986) Polymer Engineering & Science, 26: 1419
9. Ebdon JR (1984) Macromolecular Chemistry 3: 175
10. Simmons WW, Zanger M (1973) The Sadtler guide to the NMR spectra of polymers, Sadtler Research Lab Inc, Philadelphia
11. Pham QT, Petiaud R, Waton H (1982) Proton & carbon NMR spectra of polymers, Wiley, New York, vol 1
12. Pham QT, Petiaud R, Waton H (1983) Proton & carbon NMR spectra of polymers, Wiley, New York, vol 2
13. Pham QT, Petiaud R, Llauro MF, Waton H (1985) Proton & carbon NMR spectra of polymers, Wiley, New York, vol 3
14. Field LD, Sternhall S (1989) Analytical NMR, Wiley
15. Müllen K, Pregosin PS (1976) Fourier transform NMR techniques: a practical approach, Academic Press, New York, London
16. Shaw D (1987) Fourier transform NMR spectroscopy, 2nd ed, Elsevier Science, New York
17. Schafer J, Natusch DFS (1972) Macromolecules 5: 416
18. Axelson DE, Levy GC, Mandelkern L (1979) Macromolecules 12: 41
19. Levy G et al. Lab One NMR 1 program, NMR Inc, 719 East Genessee St, Syracuse, New York 13210
20. Nardin R, Vincendon M (1986) Macromolecules 19: 2452
21. Bruch MD, Bonested JK (1986) Macromolecules 19: 1622
22. Bruch MD, Payne WG (1986) Macromolecules 19: 2712
23. Crowther MW, Szenverenyi NM, Levy GC (1986) Macromolecules 19: 1333
24. Moad G, Rizzardo E, Solomon DH, Johns SR, Willing RI (1986) Macromolecules 19: 2494
25. Patt SL, Shoolery JN (1982) J. Mag. Reson., 46: 535
26. Doddrell DM, Pegg DT, Bendall MR (1982) Chem. Phys. 77: 2745
27. Heatley F, Luo Y-Z, Ding JF, Mobb RH, Booth C (1988) Macromolecules 21: 2713
28. Fox JJ, Martin AE (1940) Proc. Roy. Soc., A175: 208
29. Cross LH, Richards RB, Willis HA (1950) Discussions Faraday Soc., 9: 235
30. Willbourn AH (1959) J. Poly Sci., 34: 569
31. Cudby MEA, Bunn A (1976) Polymer 17: 345
32. Harma T, Suzuki T, Kasaka K (1975) Kobunishi Ronbunshi 32(2): 91
33. Bovey FA, Schilling FC, McCracken FL, Wagner HL (1976) Macromolecules 9: 363
34. Nishioka A, Mukai Y, Ouchi M, Imani M (1980) Bunseki Kagaku 29(11): 774
35. Randall JC (1978) J. Appl. Polym. Sci., 22: 585
36. Usani T, Gotoh Y, Takayama S (1986) Macromolecules 19: 2722

37. Hay JN, Mills PJ, Ognfanovic R (1986) Polymer 27: 677
38. Grant DM, Paul EG (1964) J. Amer. Chem. Soc., 86: 2984
39. Lindeman LP, Adams JQ (1971) Anal. Chem., 43: 1245
40. Randall JC (1973) J. Polym. Sci. Polym. Phys., Ed. 11: 275
41. Heatley F, Salovey R, Bovey FA (1969) Macromolecules 2: 619
42. Mitani K, Suzuki T, Matsuo A, Takegami Y (1974) J. Poly. Sci. Polym. Chem. Ed 12: 771
43. Stehling FC, Knox JR (1975) Macromolecules 8: 595
44. Ferguson RC (1971) Macromolecules 4: 324
45. Kakugo M, Miyatake T, Naito Y, Mizunuma K (1988) Macromolecules 21: 314
46. Hayashi T, Inoue Y, Chujo R, Asakura T (1988) Polymer 29: 138
47. Cheng HN, Lee GH (1987) Macromolecules 20: 436
48. Inoue Y, Itabashi Y, Chujo R (1984) Polymer 25: 1640
49. Cheng HN (1985) Polym. Bull. (Berlin) 14: 347
50. Asahura T, Nichiyama Y, Doi Y (1987) Macromolecules 20: 616
51. Hashimoto M, Miyaka T, Masuda Y, Ya Mayata F (1987) Nippon Gornu Kyokaishi, 60: 38
52. Bunn A, Cudby MEA (1976) Polymer 17: 548
53. Randall JC (1978) Macromolecules 11: 592
54. Bunn A, Cudby MEA, Randall JC (1980) Polymer 21: 117
55. Ray GJ, Johnson PE, Knox JR (1977) Macromolecules 773: 10
56. Carmen CJ, Wilkes CE (1971) Rubber Chem. Techno 44: 781
57. Randall JC (1978) Macromolecules 11: 33
58. Okada T, Hashimoto, Ikushige T (1980) Nippon Kagaku Kaishi 1: 74
59. Wu TK, Ovenall DW, Reddy GS (1974) J. Polym. Sci. Polym. Phys. Ed. 12: 901
60. Ketels H, Beulen J, Van der Velden G (1988) Macromolecules 21: 2032
61. Zambelli A, Gatti G (1978) Macromolecules 11: 485
62. Harwood HJ (1968) J. Polym. Sci. Part C, 25: 37
63. Allen PEM, Fisher MC (1985) Eur. Poly. J., 21: 201
64. Ferguson RC, Ovenall DW (1985) Polym. Prep. (Amer. Chem. Soc. Dev. Polym. Chem.), 26: 182
65. Bandermann F, Speikamp HD (1985) Makromol, Chem. Rapid Commun., 6: 335
66. Kricheldorf HR, Schwartz M (1983) Makromol Chemie 184: 475
67. Hvilsted S, Jorgensen NU (1983) Polymer Bulletin 10: 236
68. Kricheldorf HR (1978) Makromol Chemie 179: 2133
69. Newmark RA (1980) J. Poly. Sci. Poly. Chem. Ed. 18: 559
70. Russell GA, Henrichs PM, Hewitt JM, Grashol HR, Sandu MA (1981) Macromolecules 14: 1764
71. Kricheldorf HR, Droscher M, Hull WE (1981) Polymer Bulletin 4: 547
72. Kricheldorf HR, Hull WE (1978) Poly. Sci. Polym. Chem. Ed. 16: 2253
73. Kricheldorf HR, Hull WE (1977) J. Macromol. Sci. Chem. A11(12): 2281
74. Kricheldorf HR, Muhilhaupt R (1977) Angew Makromol Chem., 65: 169
75. Tran N-M, Djodeyre E, Carriere FJ, Sekiguchi E (1980) Makromol Chem. Rapid Commun., 1: 507
76. Kricheldorf HR, Schilling G (1976) Makromol Chemie 177: 607
77. Schilling G, Kricheldorf HR (1975) Makromol Chemie 176: 3341
78. Kricheldorf HR, Schilling G (1978) Makromol Chemie 179: 2267
79. Kricheldorf HR (1978) Proceedings of Eur. Conf. on NMR of Macromolecules
80. Kricheldorf HR, Leppert E, Schilling G (1975) Makromol Chemie 176: 1629
81. Kricheldorf HR, Kaschig J (1978) J. Eur. Polym. J., 14: 923
82. Kricheldorf HR, Mülhaust R (1980) J. Macromol Sci. Chem., A14(7): 977
83. Cudby MEA, Feasey RG, Jennings BE, Jones MEB, Rose JB (1965) Polymer 6: 589
84. Ger. Offen. (1978) 2803873, 1978 Chem. Abst. 1978, 89, 147398g
85. Attwood TE, Dawson PC, Freeman JL, Hoy LRJ, Rose JB, Staniland PA (1981) Polymer 22: 1096
86. Lyle GD, Senger JS, Chen DH, Kilic S, Wu SD, Moherty DK, McGrath JE (1989) Polymer 30: 978
87. Bunn A (1988) Brit. Polym. J., 20: 307
88. Abraham RJ, Haworth IS, Bunn A, Hearmon RA (1988) Polymer 27: 1110
89. Abraham RJ, Bunn A, Haworth IS, Hearmon RA (1989) Polymer 30: 1969
90. Abraham RJ, Bunn A, Haworth IS, Hearmon RA (1990) Polymer 31: 728
91. Attwood TE, Newton AB, Rose JB (1972) Br. Polym. J. 4: 391
92. Heatley F (personal communication)

93. Starnes WH, Schilling FC, Abbas KB, Cais RE, Bovey FA (1979) Macromolecules 12: 556
94. Starnes WH, Schilling FC, Abbas KB, Cais RE, Bovey FA (1979) Polym. Prepr., Am. Chem. Soc., Div. Polym. Chem. 20: 653
95. Starnes WH, Schilling FC, Plitz IM, Cais RE, Freed DJ, Bovey FA (1980) Preprints, 3rd International Symp. on PVC, Cleveland
96. Starnes WH (1981) Development in polymer degradation, 3, S. 135
97. Starnes WH, Schilling FC, Plitz IM, Cais RE, Bovey FA (1981) Polym. Bull. 4: 555
98. Starnes WH (1985) Pure & Appl. Chem., 57: 1001
99. Bovey FA (1962) J. Poly. Sci. 62: 197
100. Ishigura K, Watanabe S, Tabata Y, Oshima K (1980) Macromolecules 13: 1630
101. Katritzky AR, Smith A, Weiss DE (1974) J. Chem. Soc. Perkin Trans., 2: 1547
102. Harwood HJ, Ritchey WN (1965) J. Poly. Sci. B., 3: 419
103. Kale LT, O'Driscoll KF, Dinan FJ, Uebel JJ (1986) J. Poly. Sci. Polym. Chem. Ed. 24: 3145
104. Brar AS, Kapur GS (1988) Polymer Journal 20: 371
105. Hirai H, Koinuma H, Tanabe T, Tacheuchi K (1979) J. Poly. Sci. Poly. Chem. Ed. 17: 1339
106. Koinuma H, Tanabe T, Hirai H (1981) Macromolecules 14: 883
107. Llauro MF, Pichot C, Ramitez W, Guillot J (1986) Poly. Mater. Sci. Eng. 54: 61
108. Llauro-Darricardes MF, Pichot C, Guillo J, Rios GL, Gruz EMA, Guzman CC (1986) Polymer 27: 889
109. Wilson CW, Santee ER (1965) J. Poly. Sci. Pt. C, 8: 97
110. Pianca M, Bonardelli P, Tato M, Cirillo G, Moggi G (1987) Polymer 28: 224
111. Weigert FJ (1971) Org. Magn. Resonance 3(3): 373
112. Ishigure K, Ohashi H, Tabata Y, Oshima K (1977) Macromolecules 10: 567
113. Braeme EG (1974) Nucl. Magn. Resn. Spectrosc. Nucl. other than protons, Wiley, New York
114. Ishigure K (1976) Buneski Kiki 14: 610
115. Devaux J, Daoust D, Legras R, Dereppe JM, Nield E (1989) Polymer 30: 161
116. Williams DAR, Dyke R, Malcomsen S, Massey S (1983) J. Cell. Plast., 19: 121
117. Noshiro M, Jitsuigiri Y, Shigeyuki K, Tadishi S, In: Ashida K, Frisch KC (eds) International progress in urethanes, vol 5 (in press)
118. Sauer W, Kimmer W (1980) Plaste kautsch 27: 188
119. Pillon LZ, Utracki LA (1984) Polymer Eng. & Science 24: 1300
120. Devaux J, Godard P, Mercier JP (1982) J. Poly. Sci., Polymer Phys. Ed. 20: 1875
121. Devaux J, Godard P, Mercier JP, Toillaux R, Dereppe JM (1982) J. Poly. Sci. Polymer Phys. Ed. 20: 1881
122. Devaux J, Godard P, Mercier JP (1982) J. Poly. Sci. Polymer Phys. Ed. 20: 1895
123. Godard P, Dekoninck JM, Devlesaver V, Devaux J (1986) J. Sci. Part A, Polymer Chemistry 24: 3301
124. Godard P, Dekoninck JM, Develsaver V, Devaux J (1986) J. Poly. Sci. Part A, Polymer Chemistry 24: 3315

Author Index Volumes 21–29

MIX
Papier aus verantwortungsvollen Quellen
Paper from responsible sources
FSC® C105338

If you have any concerns about our products,
you can contact us on
ProductSafety@springernature.com

In case Publisher is established outside the EU,
the EU authorized representative is:
Springer Nature Customer Service Center GmbH
Europaplatz 3, 69115 Heidelberg, Germany

Printed by Libri Plureos GmbH
in Hamburg, Germany